A BRIEF HISTORY OF THE PHILOSOPHY OF TIME

A BRIEF HISTORY OF THE
PHILOSOPHY OF TIME

Adrian Bardon

OXFORD
UNIVERSITY PRESS

OXFORD

UNIVERSITY PRESS

Oxford University Press is a department of the University of Oxford.
It furthers the University's objective of excellence in research, scholarship,
and education by publishing worldwide.

Oxford   New York
Auckland   Cape Town   Dar es Salaam   Hong Kong   Karachi
Kuala Lumpur   Madrid   Melbourne   Mexico City   Nairobi
New Delhi   Shanghai   Taipei   Toronto

With offices in
Argentina   Austria   Brazil   Chile   Czech Republic   France   Greece
Guatemala   Hungary   Italy   Japan   Poland   Portugal   Singapore
South Korea   Switzerland   Thailand   Turkey   Ukraine   Vietnam

Oxford is a registered trademark of Oxford University Press
in the UK and certain other
countries.

Published in the United States of America by
Oxford University Press
198 Madison Avenue, New York, NY 10016

Library of Congress Cataloging-in-Publication
Data Bardon, Adrian.
A brief history of the philosophy of time/Adrian Bardon.
pages cm
ISBN 978-0-19-997645-4 (alk. paper)—ISBN 978-0-19-930108-9 (pbk. : alk. paper)—
ISBN 978-0-19-997773-4 (updf)—ISBN 978-0-19-997774-1 (epub)
1. Time. I. Title.
BD638.B335 2013 115—dc23
2012045782

1 3 5 7 9 8 6 4 2
Printed in the United States of America
on acid-free paper

For Janna, Zev, and Max

CONTENTS

ACKNOWLEDGMENTS

I am very grateful for the extensive comments given on drafts of this work by Heather Dyke, Craig Callender, L. Nathan Oaklander, Barry Dainton, Yuri Balashov, and Eric Schliesser; I am also grateful for the comments by six anonymous reviewers selected by Oxford University Press, and Greg Cook generously clarified a number of issues.

I used this book as a text in several recent editions of my Philosophy of Space and Time course; I am grateful for the responses I received from my students.

Marcia Underwood kindly rendered the illustrations in digital form; her creativity and design talents improved them in many ways.

My wife, Janna Levin, read the manuscript more than once, looking for errors and providing helpful comments.

A BRIEF HISTORY OF THE PHILOSOPHY OF TIME

What Does It Mean to Ask, "What Is Time?"

What makes time so elusive? Time couldn't be a more familiar and fundamental part of our existence—and yet, as soon as we really start thinking about it, we find that there is no subject more mysterious and ineffable. 'Ineffable' is a particularly good way to put it: It means 'beyond words.' It is difficult to get started in thinking about time, because it is difficult even to put our thoughts about time into words.

The basic problem has been under intense consideration throughout recorded history. There are two essential facts about time that most will agree on. First, we think of events as arrayed in a sort of order, where what is happening depends on where we are in that order. Second, we think of events as coming to be and passing away, as undergoing change over, or in, time. (Roughly speaking, we use calendars to track this first aspect of time and clocks for the second.) But these two characteristics seem to be in tension: If events are arrayed in an order, then how can we also say that they come to be and pass away? Is the passage of time real, or is it merely a subjective aspect of our experience? What is it for an event to be 'in' time in the first place? Upon reflection, it is very difficult to explain just what a temporal description of the world really amounts to.

This fundamental conundrum gives rise to a number of significant subsidiary questions. What is the nature of our experience of time? What gives time its direction? Is travel in time possible? Is the future unwritten, and do our choices matter? Did time begin, and, if so, how?

This book concerns the *philosophy* of time. One might well wonder how a philosophical approach to time is different from a scientific, psychological, sociological, literary, or other approach to the subject. Answering this question requires that we briefly examine what philosophy is.

To be honest, philosophers generally dread being asked to explain what philosophy is. Part of the problem is that philosophy is more of an activity—the activity of philosophical thinking—than a subject matter, so it is easier to demonstrate than to define. Unlike physics, mathematics, literary studies, religious studies, or just about any other field of investigation, philosophy does not have its own, unique subject matter: A given philosophical investigation might, for example, concern itself with the subject matter of science, or math, or art, or religion. Philosophy is really distinguished by the *kinds* of questions it asks. Philosophers ask foundational questions—questions about, say, science: What is a scientific explanation? What is causation? What is the proper domain for empirical study? Philosophers ask questions about art: What is beauty? What counts as a work of art?

There is an unwarranted prejudice that philosophers like to dither around and ponder unanswerable questions. Nothing could be further from the truth, at least as far as contemporary academic philosophy is concerned. Thinking about philosophical questions is not viewed by philosophers as some sort of meditation, with no real endpoint. Philosophers deal in tough, abstract questions, but they shun unanswerable ones like the plague. Indeed, distinguishing between questions that are hard to answer and questions that are meaningless or otherwise poorly formed is a big part of the philosophical enterprise. The inherent difficulty of philosophical questions can make progress very slow, and this may be confused with a lack of progress.

To get a better grasp of what time is all about, philosophers have two main jobs to do: figure out exactly what questions to ask, and then figure out how to answer them. The first of these jobs is often the tougher one, and is commonly the main task in serious philosophical work.

In understanding the question "What is time?" we start by trying to zero in on our target. Figuring out what you are asking when asking about time is less than straightforward. In ordinary discourse, we employ temporal terms, like "past," "present," and "future," without thinking much about what they mean. In describing the world, natural scientists tend to presume an understanding of temporal concepts, like temporal measurement, succession in time, or the earlier/later relation, in their accounts. Before we can formulate questions about time, we need to look carefully at what our notions about time include, and what facts and concepts we take for granted in both colloquial and scientific discourse.

Time certainly has something to do with measurement. This doesn't tell us much so far, because what time measures is duration, and duration is a temporal concept. Time can also be thought of as a coordinate system. Events are located 'in' time; they have a fixed temporal position relative to each other. This means something different from having a different spatial position, or a different position on a number line—but how, exactly? Finally, time has something to do with change. Again, this is just a starting point, because it is very tough to see how we could understand what *change* is without understanding what time is: Change involves something having different properties at different *times*. We also speak of change 'of' time: The future, we sometimes say, approaches, and the past recedes. But is this a real phenomenon or a metaphor for something else?

Then there is the problem of methodology. Philosophical questions are philosophical precisely because they demand unusual methods. Some ordinary questions can be answered by appeal to authority (e.g., by consulting a professional or looking them up in a book); other kinds of questions are answered by experimentation, observation, and inductive inference. Philosophers specialize in tackling precisely those questions that are not amenable to these everyday ways of finding things out. Philosophical methods involve innovative uses of reason and logic; a big part of any philosophical project is to figure out how to understand and address the issue in question using these tools.

There are silly, meaningless questions, and there are tough and abstract, yet answerable, questions. Questions about time, I believe, fall into the latter category. Time is a difficult and puzzling matter because questions about time tend to be questions of the tricky, philosophical sort. Asking "Is time real?" is a fundamentally different enterprise from asking whether, say, three-toed sloths are real. We know what the latter question means. We also know how to go about finding the answer: Head to Central America and other places where sloths are known to congregate, find all the sloths we can, and count their toes. We know what would constitute success or failure in finding three-toed sloths: Either we find some or, after a careful and thorough search, we don't. If the former occurs, then the problem is solved. If the latter, though we may not be absolutely certain that they do not exist, we can reasonably conclude that they do not. In contrast, figuring out whether time is real is a whole different ballgame. It is not something we are going to uncover just by looking around.

Historically, philosophers have had a particular focus on questions about knowledge and reality that are not susceptible to more mundane methods of investigation. Take the case of numbers. Is the number *seven*, for example, itself a real thing? Obviously, it is not a material item like a rock or tree or sloth. But we talk about it and solve problems with it. So it is not material, but neither is it fictional, like Sherlock Holmes or the Loch Ness monster. And how do we know that seven plus five equals twelve? We know this is true, but not in the way that we know whether sloths exist. Philosophers try to come up with ways of answering questions like these; but because of the tricky nature of the issues involved, they first have to come up with ways of grasping the meaning of these questions. And doing this tends to require tackling some even more fundamental questions, like what is it for something to be *real*, and what is it for us to know that something is *true*?

Next, consider another classic preoccupation of philosophy: moral facts. Upon examination, a murder scene may reveal a body, a bloody knife, even fingerprints. But the moral viciousness of the act of murder, no matter how closely we look, is not something we actually *see* in addition to these other elements. Moral facts, such as the fact that murder is wrong, are things we like to think we can know, but it is awfully tricky to explain just what sort of facts they are and how it is we know them. Figuring out how to locate these facts, and what would constitute success in finding them, is a nontrivial part of the process.

Take color as yet another example. We know that color varies according to the observer, as well as the lighting conditions. So, do things have a true color or not? Is color a real property in the world at all? Should color just be thought of as a disposition on the part of objects with certain other characteristics to give rise to visual experiences of a certain kind? These are philosophical questions about color. Note the relationship between these questions and a scientific understanding of color. These questions begin with the scientific description of the situation: Objects reflect light at a certain wavelength, and our brains are so organized that they typically register a visual sensation of a corresponding sort in variable but predictable ways. This basic picture is not in dispute. What are in dispute are deeper and more abstract questions like "What is it for something to be a property?" And "Where do we draw the line between objective and subjective?" These philosophical questions about color seek to deepen our understanding of the situation by pushing the subject further, into issues of how best to think about what we have learned from the science of color perception.

These examples may help us understand why we find time so weird and ineffable. The question as to whether time is real may be more like the question as to whether numbers, moral facts, or colors are real than the question as to whether sloths are real. We are badly in need of help when thinking about time, because our questions about time need clarification, and proper methods for answering these questions have to be settled upon. Fortunately, these are precisely the things that philosophers are especially good at.

What are we looking for when looking for an explanation of time? How do we know when we have the answer? Can reason and logic alone provide substantive answers? What about the empirical sciences? What is the relationship between our experience of time and time as described by the empirical sciences? How is understanding our experience of time pertinent to understanding time itself? These are philosophical questions, and in fact the history of the philosophy of time gives us lots of reasons to think that we can actually make progress on answering them.

Talented thinkers have been working on these problems for several thousand years; that is good news for us, because their having done so much of the groundwork makes things a lot clearer. Over the centuries, theories about the nature of time have resolved into three main categories: **idealism, realism,** and **relationism**. Idealists believe that time is a merely subjective matter, and nothing in reality corresponds to it. Realists maintain that time is a real thing, a kind of underlying matrix for events. Relationists take something of a middle path; they believe that time is just a way of relating events to each other, but the relations it describes are real.

Ancient Greek scholars were divided among these three basic views about time. It is with this early dispute that we begin.

Chapter 1

Time and Change

We measure weight with a scale and temperature with a thermometer. When we measure time—say, with a clock—what is it that we measure? To some scholars in the ancient world, the answer was that what we call time is simply the measure of change. What is real is a changing universe; time is derived from, and is used to track, regular changes and motions. The leading proponent of this view was the celebrated Greek philosopher Aristotle, who was both Plato's most accomplished student and himself a teacher of Alexander the Great. His philosophical opponents on this issue were Parmenides and Zeno, who denied the reality of change—and thus denied the reality of time as well.

THE ELEATICS

There was a remarkably high level of scholarly activity among the ancient peoples of the Mediterranean region around 2,500 years ago. In addition to areas of study ranging from mathematics to politics, there were a number of different schools of metaphysics (meaning schools specializing in the philosophical investigation into the nature of reality itself). One approach that many of these schools had in common was a certain emphasis on the divide between appearance and reality, and how mere appearances can be deceiving. One such influential school of thought—known as the "Eleatic" school because its proponents came from the Greek colony of Elea on the Italian coast—actually denied the reality of change. As (they would also maintain) there is no time without change, their denial of the reality of change is also a denial of the reality of time. As odd as it may sound to propose that there is no such thing as change, this theory's proponents had some surprisingly compelling things to say in favor of this claim.

The Eleatic school of thought included two of the greatest early philosophers of nature, Parmenides and Zeno. We think that Parmenides was born around 515 BCE and lived to be at least sixty-five years old. Zeno, his student and ally, was about twenty-five years younger. The only records we have of their work are snippets quoted or discussed in the writings of other ancient and medieval scholars. In their time, though, they were very well known. Their view of reality stood in direct opposition to that of their contemporary Heraclitus, who argued that reality is characterized by unending change, with nothing constant in the world. He claimed that our awareness is always changing, and what we experience is always undergoing change—even if we don't realize it, as when we identify a river as the same river, even though the water making up the flowing river is constantly replaced. The Eleatics, in contrast, shared one central belief that might seem radical: that all change is an illusion—that, in reality, all the world is an unchanging, timeless unity. In philosophical terms, then, they were **idealists** about time: Time is merely a kind of idea in the mind, rather than a thing that can genuinely be attributed to nature. What could be said in favor of such a counterintuitive notion? Quite a bit, as it turns out.

ZENO'S PARADOXES

The Eleatic denial of change is the point of Zeno's famous paradoxes of motion. These paradoxes have fascinated scholars for thousands of years. Zeno came up with an unknown total number of these; they are variously examined by a number of ancient scholars, particularly Aristotle. The three paradoxes most often discussed, and most closely related to Parmenides' shared assumptions about change, are known as "The Dichotomy," "Achilles and the Tortoise," and "The Arrow."

1. The Dichotomy

Suppose someone (call her Atalanta) intends to walk, at a steady pace, to the grocery store down the block. In order to arrive at the store, Atalanta must first cover half the distance to the store. This leaves her with a certain distance yet to cover. In order to reach the store from there, she must again cover half the remaining distance (one-quarter of the total distance). Having accomplished this, she must cover half the remaining distance (one-eighth the total), and so on. This means that Atalanta, in order to reach the store, must cover an infinite number of finite distances (see figure 1.1). She could not do this in a finite amount of time; therefore she would never get to the store.... Of course, people successfully reach their destinations all the time, thus the paradox.

2. Achilles and the Tortoise

Zeno's second paradox of motion makes a point similar to the first. He imagines epic anti-hero Achilles—apparently known for his foot speed until his unfortunate heel injury—racing against a much slower

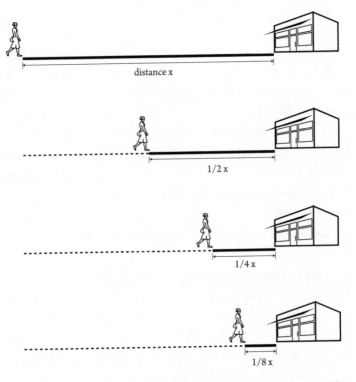

distance x

1/2 x

1/4 x

1/8 x

Figure 1.1. And so on, ad infinitum. It seems that Atalanta has an infinite number of finite distances to traverse before she can reach the store.

tortoise. Suppose Achilles gives the tortoise a head start. For Achilles to catch up to the tortoise, he must first reach the point (call it point A) where the tortoise was when Achilles started running. By that time, the tortoise will have moved forward to a farther point B and will still be progressing slowly. In order to catch up to the tortoise, Achilles must first get to point B, and the same for point C, and so on. Similarly to the first paradox, it appears that catching the tortoise will take an infinite number of steps—a series of steps we know will have no end—each of which will require a finite (albeit ever-diminishing) amount of time to accomplish. Could such a series of tasks ever be completed?

3. *The Arrow*

Zeno's third paradox makes a different kind of argument. Take an arrow flying through the air. Is it really in motion? It can't be, according to Zeno, for the following reason: Whether or not something is in motion should be a fact about that thing *now*, not a fact about it in the past (i.e., that it was somewhere else) or a fact about it in the future (i.e., that it will be somewhere else). But at any given instant, the arrow only occupies a space equal to its own size, and when something occupies only a space equal to its own size, it is at rest. Therefore the arrow cannot be in motion at any time. Because the same reasoning would apply to any object in motion, motion is impossible.

Zeno concludes that, even though it appears for all the world as if things move and change, reason and logic rule out the possibility of motion. His resolution of these paradoxes is that movement, and change in general, is an illusion. The true nature of the world (i.e., its unchanging perfection) is revealed to us when we ignore our sense experience and rely on reason alone.

ARISTOTLE'S ANSWER TO ZENO

We know about Zeno mainly through Aristotle, who resisted the Eleatic attack on the reality of change. For Aristotle, change is real: Objects move from one place to another, seasons change, ice melts, and so on. He saw a relationship between time and change, but the nature of that relationship requires some explanation.

Plato, Aristotle's teacher, had expressed sympathy for a kind of naïve realism about time that simply identified time with change. In Plato's dialogue *Timaeus*, the character of Timaeus identifies time itself with the motion of the heavenly bodies (the Sun, the Moon, and the planets). He goes on to claim that time would come to an end if these bodies would cease their orbits. On this view, then, time is related to change in the sense of simply being identified with a particular set of motions. It is not clear what Plato himself really thought, but he does describe Timaeus' story as a "likely account" of time. Aristotle finds this doctrine unpalatable, pointing out that, even if the heavenly bodies stopped dead in their tracks, time would still be passing as long as some other things were in motion.

Nor can we simply identify time with motion or change in general, he continues. He notes that change is a contingent and local phenomenon, whereas we think of time as passing equally for everything everywhere, no matter what is going on in the immediate vicinity. Further, although changes can be slower or faster, time cannot; "slow" and "fast" are defined in terms of time, not vice-versa. So time cannot literally be change.

Rather, for Aristotle, the relationship between change and time is not one of identity, but more like the relationship between the thing measured and the means of measuring it. Time is not a process: It is just a kind of "number" or unit that can be used to describe processes in nature, analogous to the way ordinary numbers can be used to count things. In his words, "time is the number of change with respect to before and after." As with other abstract quantities, time is a kind of system that captures something real about nature without really being part of nature. If you see two sheep, there isn't really any 'twoness' *out there* in addition to the sheep, but that is not to say that there aren't really two of them. We can truly say that the Sun is brighter than the Moon, and that an elephant is bigger than a mouse, even though 'brighter' and 'bigger,' as mere relations, aren't really things in the world like suns, moons, elephants, and mice. In the same way, we can say that a performance of *Hamlet* really is two hours longer than a TV cartoon show, even though there are no hours *out there* to count. What we do is use regular motions, like the orbit of the Earth or Moon, or the ticking of a clock (in Aristotle's time, more likely the dripping of a water clock), as units of duration, which in turn can be used to count, order, or measure other durations, motions, or changes. Time itself exists only in the sense that it is a unit system used to count, order, or measure such things. Yet this doesn't mean that we can't make perfectly accurate statements about the time and temporal order of events. This view about time is a variant of what is called **relationism**, in that it treats time as a way of thinking about how events can be objectively related to each other.

Aristotle's theory allows an answer to Zeno that preserves change as a genuine aspect of reality. He thinks that Zeno's paradoxes rest on a confusion between time and what it measures. For Aristotle, time is a unit of measurement used to describe changes; as such, it is a quantity that belongs more in the mathematical than the material realm. Zeno's paradoxes, he argued, are dissolved by the fact that concepts like infinity, composition, and infinite divisibility work differently when you are talking about (merely) mathematical quantities in a stipulated mathematical realm. First, consider the paradox of "The Dichotomy" and the similar paradox of "Achilles and the Tortoise." On Aristotle's conception, time is like the number line, in that any segment of either of these is, potentially, infinitely divisible. You can take the 'space' between, say, the integers two and three and divide it in half, then in quarters, then in eighths, and so on indefinitely. He argues that what this really means is that, although you can identify infinitely many distinct points between the numbers two and three (corresponding to each possible division of that segment of the number line) and infinitely many distinct points within any smaller division of any subdivision of that distance, the distance between two and three is not (and couldn't be) *composed* of those points in the way material objects are composed of smaller parts. The stipulated rules for an abstract mathematical realm mustn't be confused with the actual rules for physical reality. A point in Euclidean geometry has zero length, and no number of things with zero length can, in the material world, be added together to form something of finite length. Geometrical points are really just abstract boundaries on merely potential subdivisions. Similarly, Aristotle argues, for instants in time. A length of time—say, the time it takes to walk to the grocery store—is not actually composed, in the material, physical sense, of an infinite number of smaller, finite lengths of time. Zeno's first two paradoxes of motion rest on supposing that the time it takes to get from one place to another is actually composed of an infinite number of finite lengths of time; if it were, then he would be right that any motion requires the completion of an infinite number of distinct tasks. Aristotle concludes that Zeno's paradoxes rest on confusing time (an abstract unit system) with change (a real phenomenon that can be measured in terms of time units). The existence of these false paradoxes, he would like to say, just goes to show that he has the right analysis of time.

Has Aristotle thus solved the problem posed by the first two paradoxes? Some have claimed that Zeno's problem was only really answered with some new mathematical ideas in the nineteenth century. Contemporary mathematics adds a concept that, if permitted to describe change itself, would more directly resolve the paradoxes. The concept of a **limit** allows for an infinite number of finite quantities to add up to a finite sum, in cases like the one in "The Dichotomy": In modern calculus, the sum of $(1/2 + 1/4 + 1/8 + 1/16 \dots)$ is said to approach, or converge on, one. This quantity is the "limit" of that particular process of addition. (In contrast, the sum of the series $[1 + 2 + 3 + 4...]$ has no limit.) If we accept that completing a motion is like reaching a limit, then neither Atalanta nor Achilles need do the impossible in order to accomplish their tasks. In one respect, this distinction meshes nicely with Aristotle's assertion that the "potential" infinity in question in "The Dichotomy" and "Achilles and the Tortoise" paradoxes is not to be confused with an "actual," extensional infinity. But where Aristotle's solution was to avoid contradiction by distinguishing between the rules for time (as a mere abstraction) and those for change (as a real phenomenon), the notion of a limit gives us the resources to describe, without contradiction, a finite sequence as actually divisible into an infinity of finite sequences. In other words, this answer doesn't depend on drawing a line between time-the-abstraction and change-the-reality the way Aristotle's does. However, this functions as a solution only if one is comfortable with the notion that one can just stipulate an answer to the paradox by simply declaring that a converging series actually converges. Can a limit be a real endpoint to a real process, or is it just a new mathematical convention that disregards the metaphysical question about time and change with which Zeno and Aristotle are struggling? Does it really help matters to say that Atalanta's progress to the store represents convergence on a limit? That wouldn't have sounded like real motion to either Zeno or Aristotle. Even if it might not be satisfying for other reasons, Aristotle's approach has the virtue of not forcing real actors to perform weird mathematical tricks.

Although Zeno's paradox of "The Arrow" appears to be making a different argument, fundamentally, Aristotle's answer to it rests on the same point, namely that "The Arrow" problem depends on the false premise that, if time is not a mere illusion, it must therefore be actually composed of instants. Zeno presumes that something in motion can only be in motion by virtue of its state at any instant. Aristotle would prefer to say that motion is motion over an interval with non-zero duration; that is why we describe motion in terms like "miles per hour" or "meters per second." By definition, an instant per se has zero duration, and so the idea of motion over an instant is incoherent: Distance traveled per zero seconds is not a rate of motion. Similarly for the notion of rest at an instant: By Zeno's reasoning, to say that something is at rest at an instant is to say that it moved zero meters *in zero seconds*. This description of rest doesn't make any sense either; rather, we should say that something at rest is moving zero meters *per second*. As with the other paradoxes, Aristotle's diagnosis is of a false paradox resting on the confusion of an abstract value (i.e., time), which is mathematically divisible into instants, with actual change, which is not literally composed of infinitesimal units of change.

The strongest virtue of Aristotle's theory is that it quickly dissolves these counterintuitive paradoxes. Further, note that he has replied to Zeno on the Eleatics' own terms (i.e., rationally). In most cases, when we want to say the world is a certain way and not otherwise, we tend to think that the best way to accomplish this is by observation and experiment. But the Eleatics' whole worldview is based on the notion that sense experience is fundamentally unreliable. Aristotle realizes that, in replying to them, it would be question-begging to rely on observation and evidence; rather, his analysis of time rests on *reasoning* about time and change alone, and so it is not open to any reply by Zeno that he is just begging the question as to our knowledge of the natural world.

PARMENIDEAN IDEALISM

Aristotle doesn't have too much trouble dispensing with Zeno's paradoxes. His argument against Parmenides, Zeno's fellow temporal idealist, is another story. Parmenides had offered a different argument altogether for the thesis that change is an illusion. Sometime in the fifth century BCE, he wrote a lengthy prose poem (existing for us only in tantalizing fragments) in which he presented a line of reasoning discounting the possibility of change—a line of reasoning that is, incidentally, particularly interesting because it is widely thought to constitute *the world's earliest surviving example of extended philosophical argumentation*:

> As yet a single tale of a way
> remains, that it is; and along this path markers are there
> very many, that *What Is* is ungenerated and deathless,
> whole and uniform, and still and perfect;
> but not ever was it, nor yet will it be, since it is now together
>   entire,
> single, continuous; for what birth will you seek of it?
> How, whence increased? From not being I shall not allow
> you to say or to think: for not to be said and not to be thought
> is it that it is not. And indeed what need could have aroused it
> later rather than before, beginning from nothing, to grow?
> Thus it must either be altogether or not at all.
> Nor ever from not being will the force of conviction allow
> something to come to be beyond it: on account of this neither
>   to be born
> nor to die has Justice allowed it, having loosed its bonds,
> but she holds it fast. And the decision about these matters lies
>   in this:
> it is or it is not; but it has in fact been decided, just as is necessary,
> to leave the one unthought and nameless (for no true
> way is it), and <it has been decided> that the one that it is indeed
>   is genuine.

And how could *What Is* be hereafter? And how might it have
been?
For if it was, it is not, nor if ever it is going to be:
thus generation is extinguished and destruction unheard of.
(Palmer, trans.)

There are several distinct points being made in this fascinating
fragment. For Parmenides, the thought of change is the thought of
something becoming something else, which necessarily involves
the thought of some thing (or state of a thing, like the fading color
of a rose) going from being future, to being present, to being past.
The thought of change, then, is bound up with the thought of the
passage of time from future to present, or present to past. It is common to say that something "awaits us in the future," or "lies in our
past." This is to treat the future and past as though they are real, the
way other places are real even though you are not there to see them.
However, in our ordinary way of thinking about time, it is also the
case that the past and the future are contrasted with the present,
precisely in that the present is thought of as real and the past and
future are not—if something is real, then it is real *now*. If future and
past events were real now, then they would be present! So to think
of what is past or future is to think of what *is not*. ["And how could
What Is be hereafter? And how might it have been? For if it was, it is
not, nor if ever it is going to be."] Thus Parmenides' conclusion that
we contradict ourselves when we describe the world temporally:
Any talk about change involves talking about the past or future as
real *and* as not real.

To this he adds that our nonsensical talk about change gets
us into strange habits and beliefs, like talking about non-present
things, which are really nothing, as though they are something ["for
not to be said and not to be thought is it that it is not"]. Also, by

being committed to change, we are effectively saying you can get something from nothing: Because the future is nothing, the coming about of any event is a case of something coming from nothing. Nothing could explain the existence and characteristics of a thing arising from, literally, nothing, and nothing could explain why it would arise then, as opposed to at some other moment ["what need could have aroused it later rather than before, beginning from nothing, to grow?"]. Real change would also suggest that things can pass out of present existence—but where would they go?

To this one could reply that there is, seemingly, another dimension to time independent of change. What about something, like an unmoving rock or table, that persists or endures, unchanged? Even as we concede that there is no change, wouldn't there still be the time that passes as a thing endures? Parmenides anticipates this objection. He replies that an enduring thing would have to have temporal parts: parts that exist now, plus parts that did exist but don't now, and parts that only will exist. But then the thing would, once again, both exist and not exist. An existing thing can't have non-existing parts ["but not ever was it, nor yet will it be, since it is now together entire"].

The moral of this story, for Parmenides, is that change is an illusion, and so the world as it appears to us cannot be real. This was what the Eleatic school was all about: the view that the world as we know it, consisting in a constantly shifting array of impermanent objects and their relations and characteristics, is a matter of mere subjective appearance. Another fragment from the poem:

> ... for nothing else <either> is or will be
> besides *What Is*, since it was just this that Fate did shackle
> to be whole and changeless; wherefore it has been named all
> things

that mortals have established, trusting them to be true,
to come to be and to perish, to be and not to be,
and to shift place and exchange bright color. (Gallop, trans.)

Insofar as sensation leads us to accept the reality of change, our senses are fundamentally deceptive. Our experience of motion, change, and the passage of time is a projection of our own limited perspective on reality. 'Time talk' is really incoherent, a fact hidden by our facility with the use of tenses and other temporal language. We see this, Parmenides would say, when we step back from what our senses and instincts are telling us and reason coldly about whether our assumptions really make sense. The world as it is in itself is a singularity: unitary, unchanging, perfect.*

As extreme as its conclusions are, the Eleatic worldview should not be dismissed too easily. What guarantee do we have that sense experience provides an accurate picture of how the world is? It is true that we would expect that our senses would evolve in such a way that we are able to interact with the world around us in a survival-enhancing way, and in turn this would seem to imply that our belief-forming faculties tend to be reliable; after all, we would have a tough time negotiating our environment if our beliefs about our surroundings were frequently false. But it is less clear that deep insight into the true nature of the world is necessary to survival, and so it should not be taken for granted that sense perception reveals the true nature of the real world in all its aspects. The beliefs we need from a survival standpoint have to do with predicting what we will experience, given what we are experiencing now, and what actions

---

* This might sound like a mystical or religious position, except for two key facts: First, the Eleatics rely for their conclusions on reason and logic, rather than faith; and second, because nothing, on their view, can come into existence, there could have been no creation of the universe.

we take. It would be possible to be very successful at such predictions while still being quite in the dark as to whether the ways we categorize and contextualize our experiences accurately reflect reality. In the absence of more sophisticated calculations, nothing could be more natural than to think the Sun revolves around the Earth; and thinking this wouldn't cause us any problems in our day-to-day efforts to survive and reproduce. Tomatoes are nutritious even if we believe they are vegetables; water is healthful and refreshing even if we deny it to be made up of H_2O, or lack a notion of molecules entirely. It would make sense for us to instinctively fear all snakes and spiders, even if only some types are dangerous. Similarly, the fact that thinking in terms of change 'works' for us doesn't necessarily mean things really change. Maybe representing things temporally is just our way of picturing the world as best we can, given a limited perspective, a bit like the way a two-dimensional painting can suggest a three-dimensional landscape.

Aristotle has an answer to Parmenides, but it is incomplete and thus less helpful than his answer to Zeno. Aristotle agrees with Parmenides' assertion that our use of "now" or "the present" to refer to more than an instant (as in "the present day," for example) is problematic from a metaphysical point of view; at any given moment, part of our day would be past and part of it would be future. Taking such loose temporal talk seriously would get us into the sort of contradiction Parmenides exploits in his argument. Aristotle thinks he can resolve the problem by distinguishing between different kinds of change. His answer focuses on Parmenides' claim that no existence can emerge from nonexistence. To this, he replies that change is not the emergence of something from nothing. All we have to do is to distinguish between the thing that changes and its configuration, its properties, or its aspects. These are what do the changing. If a person turns pale, the person does not come into being from nothing; rather, there is a persistent thing—the person—with variable attributes. So the person, in Parmenides' terms, can be both pale and not pale without herself both being and not being. It is also true that persons can come into being, but only, Aristotle would argue, by a reshaping of an existing underlying substance, just as molten bronze can be formed into a statue. The statue is not thereby created from nothing and, if the statue is melted down, the statue does not become nothing.

Unfortunately, with this, Aristotle has not addressed the main point. The key issue is whether, when we describe change of any sort (including Aristotle's change in properties), we commit ourselves to the reality of the future from which the new situation arises, and/or the reality of a past to which it passes. But the future and past can't be real, because what would then distinguish them from the present? If the future and past are not real, then nothing can pass from being future to being present to being past. If there is no passage, then there is no change. If there is no change, then there is no time. *That* was Parmenides' fundamental problem with change and the passage of time, and Aristotle didn't say anything to dispel this worry.

Consequently, although Aristotle may have a reasonable answer to Zeno, he does not really get to the heart of the Parmenidean argument. Parmenides' argument, and the temporal idealism it represents, was later revisited by Augustine of Hippo. Augustine (aka St. Augustine, the fifth-century Catholic bishop), was a North African of Berber descent. He was a careful and insightful philosopher, as well as the most important early Christian theologian. He wrote an enormous number of books, but is best known for his *Confessions*, which combines an account of his conversion to Christianity with some very sophisticated philosophical investigations into time, memory, and cosmology.

AUGUSTINE'S THEOLOGICAL IDEALISM

Augustine was well aware of the ancient debate over the reality of time, and was very concerned with understanding our relationship to time. Augustine's primary interest in time had a theological basis. He was worried about questions like "What was God doing before creating the universe?" and "If nothing existed (but God) before the universe was created, then why create it at one time rather than another?" This last question particularly troubled Augustine: If nothing existed but God before creation, then what could have happened to, or within, God that led God to decide to create the universe at that particular moment? The new impulse that would have to have arisen at that moment suggests a *change* for God; but why would an eternal and perfect being want or need to change?

Augustine's answer is to embrace a slightly revised set of arguments for idealism about time and change: Like Parmenides, he argues that time and change are subjective phenomena of human mentality.

His goal is to call our pre-critical grasp of time into question. He famously asks: "What, then, is time? I know well enough what it is, provided no one asks me. But if I am asked what it is and try to explain, I do not know." He offers the following line of reasoning:

> Of these three divisions of time [past, present, and future] then, how can two, the past and the future, *be*, when the past no longer is and the future is not yet? As for the present, if it were always present and never moved on to become the past, it would not be time but eternity. If, therefore, the present is time only because it moves on to become the past, how can we say that even the present *is*, when the reason why it *is* is that it is not to be? (Pine-Coffin, trans.)

So, similarly to Parmenides, the past and the future *are not* (now), so they are not real. Further, the present is ad infinitum analyzable into smaller and smaller durations: We can speak of the present day, but part of that day is past (and so nonexistent) and part is future (and so nonexistent). Same for the present hour, the present minute, and so on. Any time you pick out includes times that are not present; there doesn't seem to be an actual, identifiable present time.

Yet, Augustine continues, we are aware of extended periods of time, we can compare periods of time to each other, and we are able to compare the present situation to the way things were in the past:

> Nevertheless we do measure time. We cannot measure it if it is not yet in being, or if it is no longer in being, or if it has no duration, or if it has no beginning and no end. Therefore we measure neither the future nor the past nor the present nor time that is passing. Yet we do measure time.

How is our awareness of time possible, if there really is no past, present, or future for us to be aware of? This is a critical question for the temporal idealist. Augustine's answer is that time exists only in the mind. Memory, sensation, and anticipation leave impressions on us, and it is these that are measured and compared when we make judgments about the passage of time. Nothing outside the mind really persists; rather, "the mind's attention persists." Memory and anticipation give our experience its temporal dimension, not the veridical perception of something outside the mind actually enduring and undergoing changes. The difference between past and future is just the difference between memory and anticipation.

Time, then, for Augustine is a human invention, and it is not to be applied either to the universe, as it is in itself, or to God—who exists in a kind of Parmenidean timeless state that Augustine calls "eternity." This solves Augustine's theological problem, because questions about what God was doing before the universe was created, or why God would decide to create the universe at one time rather than another, are moot if the passage of time does not apply to God.

So Augustine, motivated by the desire to invalidate certain tricky theological questions, invokes Parmenidean temporal idealism as the solution. In the process, he raises some good questions about how we come to make judgments about past, present, and future, and the passage of time. Augustine did not realize, however, that his treatment of these questions raises a very difficult issue for the temporal idealist: How do we even come to have the *concept* of past, present, future, and the passage of time if we never actually experience them? Even if these judgments are false, they involve ideas that must have come from somewhere. The problem is that Augustine's account depends on the employment of temporal concepts *without explaining where they come from in the first place.* Augustine talks about memory and anticipation, and how they create a kind of metaphorical 'extension' of the mind that stands in for temporal extension, leading to a confusion about the objectivity of time. A memory is, by definition, a representation of the past. But what gives memory the meaning it has for us? How do we recognize a memory as a memory, as opposed to, say, a product of sensation or imagination, without knowing what it means for something to be in the past? Parmenides and Augustine agree that the past does not exist, so we never actually experience the past. And it is similarly so for the future. If the idealists are right, how does the notion of the past or future get started in the first place? In the next chapter, we examine a few different ways one might go about explaining from where these supposedly false concepts come.

WORKS CITED IN THIS CHAPTER

Aristotle. *Physics.*

Augustine. *Confessions,* trans. by R. S. Pine-Coffin (London: Penguin Books, 1961).

Gallop, David. *Parmenides of Elea* (Toronto: University of Toronto Press, 1984).

Palmer, John. *Parmenides and Presocratic Philosophy* (Oxford, UK: Oxford University Press, 2009).

Plato. *Timaeus.*

Other works of relevance to the issues in this chapter

Coope, Ursula. *Time for Aristotle* (Oxford, UK: Oxford University Press, 2005).

Hoy, Ronald. "Parmenides' Complete Rejection of Time," *Journal of Philosophy* 91 (1994), 573–598.

Huggett, Nick. *Space from Zeno to Einstein* (Cambridge, MA: The MIT Press, 1999).

Matthews, Gareth. *Augustine* (Oxford, UK: Blackwell, 2005).

Sorabji, Richard. "Is Time Real? Responses to an Unageing Paradox," *Proceedings of the British Academy* 68 (1982), 190–213.

Turetzky, Philip. *Time* (New York: Routledge, 1998).

Chapter 2

Idealism and Experience

We ended the last chapter wondering where the very idea of past and future comes from, given that (a) we could never experience past or future events directly, and (b) memory and anticipation only have the meaning for us that they do because we already understand their connection to the past and future. Clearly, we do have the idea of past and future, and have no problem understanding what it means to remember or anticipate. But how we accomplish this is a different story, with a lot to teach us about the nature of time itself. In the last chapter, we saw that the metaphysical question as to whether change is real was the focus in the ancient world; in the Enlightenment era of the seventeenth and eighteenth centuries, the epistemological question as to the origin of temporal concepts became a central concern in its own right.

LOCKE'S MISTAKE

The celebrated English philosopher John Locke is probably best known for his 1690 *Second Treatise of Civil Government*; this was the most important early treatment of the notion of government by 'social contract.' But he also studied the human mind, a study that, for him, was intimately tied up with his interest in political liberty. European history for the previous thousand years and more had been characterized by claims by a few—kings or religious leaders—to have privileged access to the truth; such claims were the basis for creating dogmas and traditions that could be exploited in consolidating power. In response, Locke spent many years working on a massive examination of human ideas, language, and knowledge titled *An Essay Concerning Human Understanding*. His main goal in this work was to show that all human knowledge is ultimately derived from a combination of experience and reflection upon experience—and nothing more. This is a doctrine known as **empiricism**. The connection between empiricism and liberty, for Locke, is that empiricism undermines the claim that there are truths that religious or civil authorities have some special access to that a properly thoughtful and scientifically minded individual cannot.

As a key part of this program, Locke spends several hundred pages trying to explain how even complex and abstract ideas are ultimately derived from sense experience alone. He devotes a chapter to explaining where the basic temporal ideas of succession, duration, and eternity come from. Locke was not an idealist about time: He was a temporal realist and follower of Newton, who thought that time and space were real entities in their own right (see chapter 3). But Locke recognized that, even on the realist view, time is not an actual object of experience—it can't be felt or seen or otherwise directly observed. So he sets out to explain how we get the idea of time from experience alone nonetheless. Here is what he has to say about the origin of the ideas of succession and duration:

> 'Tis evident to anyone who will but observe what passes in his own mind, that there is a train of ideas, which constantly succeed one another in his understanding, as long as he is awake. Reflection on these appearances of several ideas one after

another in our minds, is that which furnishes us with the idea of *succession*; and the distance between any parts of that succession, or between the appearance of any two ideas in our mind, is that we call *duration*.

By a "train of ideas," Locke is just referring to any sequence of ordinary thoughts and perceptions: I see my bus coming, then I notice that my nose itches, then I scratch my nose, then I think about whether I will be late for work. His proposal is that we originally get the notion of temporal succession by reflecting upon a succession of perceptions like these, either as they occur or afterward. Supposedly, this would give us a direct experience of succession.

It is important to see why this story can't be right or is at best incomplete. Locke is claiming that the idea of succession derives from a direct experience of successions of ideas. At any moment during a reflective experience of succession, wouldn't only one component of the succession be before the mind's eye? If so, then reflecting on a whole train of ideas altogether would have to involve the *reproduction*, in memory, of past ideas or experiences (see figure 2.1). The mere reproduction of some past experiences would not give rise to the idea of a succession: The components of the idea of a succession would have to be thought of as occurring at different times—otherwise, it wouldn't be a memory of a succession, as opposed to a single, complex thought of a bunch of overlaid mental contents. He has to be talking about a reflected-upon memory of a sequence thereby identified *as such*—as a memory. That is to say, he has to presume that we already understand that the sort of reflective activity going on references past experiences. A memory thought of *as a memory* involves already identifying something *as past*; but without the idea of temporal succession already in place, the notion of 'pastness' couldn't possibly mean anything to the experiencer. If

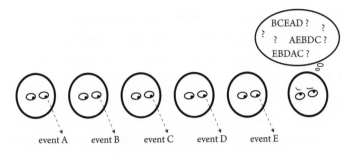

Figure 2.1. At the time of their reproduction, what reason does one have to reproduce memories in any particular order, or even to think of them *as* reproductions?

one didn't know what it is for an event to succeed another in time, one surely couldn't make sense of a present event becoming past by virtue of its being succeeded by another.

In summary, Locke does not explain how it is possible to recognize a succession of ideas or experiences as a succession without grasping, on some level, the concept of temporal succession. So his story about deriving the very idea of succession from experience can't work as it stands.*

KANTIAN IDEALISM AS A SOLUTION

The great eighteenth-century Prussian philosopher Immanuel Kant probably thought more, and more deeply, about time and our awareness thereof than anybody before him. Like Locke, he was celebrated in his own lifetime for both work in political philosophy and the study of knowledge and the mind, as well as for work in moral philosophy,

---

* There might be a story one could tell that would explain the possibility of a direct experience of succession (see the discussion of the 'phenomenological present' later in this chapter), but Locke's account is not it.

aesthetics, and many other subjects. Some of his most interesting arguments appeared in his *Critique of Pure Reason,* a massive and difficult work the main purpose of which was to resolve skeptical concerns about claims to scientific knowledge. His project was, to a large degree, inspired by what he saw as the failure of Locke and other empiricists to account for the possibility of such knowledge. Modern science draws importantly on some key concepts like material substance, causation, and space and time. Science describes the world, at a fundamental level, in terms of these concepts. To be assured we are getting things right when describing the world using these concepts, we need to explain what entitles us to apply them to our experience. Kant recognized that Locke and his fellow empiricists never gave a satisfactory explanation for the scientific use of these foundational notions.

Kant's answer required a reexamination of the origin of abstract ideas (in particular, those foundational ideas of material substance, cause and effect, and time and space) and the justification of scientific generalizations about the world employing these ideas. In turn, this required starting over at the beginning with a new investigation into how the mind gets going in formulating ideas about what is going on around it. His insight was that the key to understanding cognition was an understanding of the most fundamental cognitive achievement— the one that makes all coherent experience possible: the interpretation of one's own experience in terms of time. He found that Locke's problem with the origin of the idea of succession points the way to a solution: Time is a mere form of sensible experience, and the ordering of experiences in time derives from the mind's own imposition of order on experience. In other words, temporal idealism (Kant's version of it, anyway) is the *solution* to the problem of skepticism about the possibility of scientific knowledge.

Early on in his *Critique of Pure Reason,* Kant lays out his idealist position on space and time. He distinguishes between what we find in experience and the nature of our experience itself. Space and time are not things in themselves, he says. Rather, they are mere "forms of experience" (i.e., our manner of registering perceptions and experiences is spatial/temporal). Instead of focusing on the nouns "time" and "space," it would be more on target to focus on the adverbial application of these concepts: In other words, we don't experience things *in time and space* so much as experience them *temporally and spatially*. Kant's key finding against Locke is that there is no way even to think about experience except temporally, just as there is no way to think about material objects except spatially. Locke has it backward when he proposes that the idea of time is derived from experience; rather, experience presupposes time.

Just like Parmenides and Augustine, Kant concludes that reality is itself atemporal.* Is such a thing (i.e., the atemporality of reality) even thinkable? Yes and no, Kant would say. His theory explains imagination's inevitable failure in this regard: Because this way of experiencing things is an irreducible part of our sensibility, we literally can't imagine any other way for things to be. Yet we can intellectually come to terms with the ideality of time. In this way, the ideality of time is like the mathematical concept of infinity: We cannot imagine the infinite (we cannot call up a mental image of, say, an infinite number of apples), but we can understand what it means. Kant thinks that an atemporal reality is something we can grasp in the abstract, even if it can never mean anything to us in practical terms.

---

* Kant has been criticized for concluding that reality is atemporal just on the basis of his position that our temporal concepts can't come from an experience of real temporal relations. Indeed, the latter doesn't in itself mean that there aren't any real temporal relations. He might reply that, given his understanding of what time is, it doesn't even make sense to think of time as a concept that is applicable to a reality considered in abstraction from the way in which we experience it.

This is a summary of Kant's views on time, but we need more than this to fill out a positive account of time-awareness. It is not enough just to point out the deficiency in any empiricist approach to time. According to Kant, we apply the notion of temporal succession as a necessary condition of experience. But how does this work, if succession is not an object of experience and so must, in any particular case, be inferred? How does the inference that some event A was followed by some event B get started? The answer, for him, has to do with the concepts of *substance* and *cause*. These are concepts that are integral to the notion of a material world independent of one's mind, in that the concept of such a world is, at root, the concept of material substances interconnected by causal relations. His basic idea is that ordering one's own experiences in time is a necessary condition of any coherent thought, and the only way to get started with ordering our experiences in time is by thinking in terms of their relation to things going on around us. Kant's predecessors assumed that we start with a flow of subjective experiences, *already organized both spatially and temporally,* and infer our way to a notion of what is going on around us. He turns this completely around: He argues that there is no basis for thinking of one's experience at any moment as representative of some determinate spatially and temporally organized sequence of events, unless one's interpretation of experience is constrained by some general notions as to how the world works. We start with an innate idea (call it an instinct, if you like, or a biologically determined disposition) of how things are 'out there,' and interpret our subjective perceptions accordingly. Making sense of one's own experiences, one might say, works from the outside in, not from the inside out.

Kant's story goes as follows: Just as space and time are just forms of sensibility, the concepts of substance and cause are just rules for organizing experiences in space and time. The idea of temporal succession is innately present in our cognitive makeup, via a set of organizational principles. Thinking in terms of a succession of experiences depends on relating those experiences to successions of events outside us; and we can only do that because we already have a certain schema built in for interpreting our sensory inputs in these terms. Part of our innate information-processing scheme is that we interpret our perceptions on the presumption that we are dealing with a world of enduring items and events, interconnected by causal relations (call this the concept of *objectivity*—i.e., what it means for something to belong to a world of objects of possible experience). We are guided by this scheme into imposing a pattern on experience consistent with such a world; this pattern involves sequences of events occurring in accordance with causal rules. So this is how we come to have the idea of succession: It is a pattern corresponding to the concept of objectivity itself that we ourselves impose on sense data to make them tell a coherent story.

In processing incoming perceptual data, one of the really fundamental things we have to do is distinguish between static states of affairs and dynamic events or processes. Consider two possible experiences: the experience of walking around a house and the experience of seeing a ship leave a dock. The first involves a succession of experiences of an unchanging state of affairs; the second involves a succession of experiences of an ongoing process of change (i.e., an experience of an actual succession). In either case, the experiences present themselves successively (see figure 2.2).

Strictly speaking, the contents of the experiences themselves don't provide prima facie evidence for one interpretation (i.e., static state of affairs or dynamic event?) over another—that is, not unless you already understand the difference between a static state of affairs, experienced over time, and an ongoing event; but the ability to distinguish between the two is what is at issue. So, how are we

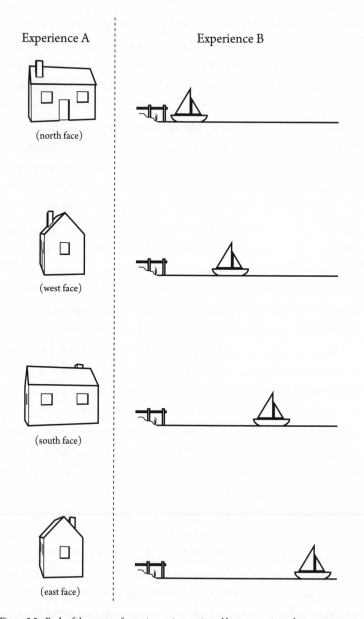

Figure 2.2. Each of these sets of experiences is constituted by a succession of perceptions, yet somehow we understand that one is the experience of an unchanging object and the other of an event in progress.

capable of this kind of distinction? Kant argues that the very notion of a difference between the two is coming from inside, so to speak: We are using one innate interpretive scheme in the one kind of case and another scheme in the other. Of course, the distinction between the two kinds of experience seems obvious, but that is the point: This is not something we have to figure out from scratch. The concept of succession according to a rule functions as a template for organizing my experiences in a certain kind of way, and I couldn't get this idea from experience (as Locke claimed), because this ability is a necessary condition of my making sense of my experience in the first place. Kant's conclusion is that the notion of rule-governed temporal succession must be innate and imposed by us on our own experience.

Kant's story is the most sophisticated idealist story so far. It has a certain plausibility in light of the difficulty empiricists have with explaining the origin of temporal concepts, insofar as time is not itself to be found in the contents of our experience. It bolsters the idealist case by accounting for the *experience* of succession (i.e., as an organizing principle of the mind) in the context of an inherently timeless reality. Kant also appreciates the fundamental importance of temporal organization to coherent experience; his account gains some credibility from the puzzle as to how we would be able to make sense of the world without some starting point—a cipher key to unlock the code—and temporal organization does seem like a reasonable candidate as just such a starting point. His account addresses Locke's experience problem, and it does so in a way that would help explain why it is so hard to put one's finger on what time is when we look for it in the world around us. In other words, it explains the ineffability of time by explaining that we have been looking for it in the wrong place: We mistakenly look for it 'out in the world,' when it is really a matter of how we organize our experience of the world.

It is important to note that Kant's theorizing falls short of proof, even if it, in principle, solves a problem generated by a simplistic empiricist approach. There remain many questions one could raise for this theory, such as how and why one is constrained to interpret a subjective sequence of perceptions according to one schema rather than another, if reality itself is not doing the constraining. Further, it remains hard to accept his conclusion that the universe is not in itself temporal at all (i.e., without us, things really do not change and time does not pass). Is there an alternative theory, compatible with realism about time, which explains what Augustine did not and Locke could not?

A REALISM-COMPATIBLE ALTERNATIVE

There are alternatives to Kantian idealism in accounting for the idea of time. It wasn't until very recently that a legitimate non-idealist alternative to Kant's solution to the problem of the origin of temporal concepts has been proposed. Again, what is needed is an explanation of how we get the idea of past and future, and change over time, from experience without having to presume an understanding of those very things. Since the nineteenth century, cognitive psychologists have been entertaining a notion that would allow for the idea of temporal succession to be derived from experience: The proposal is that our experience at any moment *directly* encompasses both the present *and* the immediate past. If so, that means that we experience succession or change directly, without having to rely on question-begging inference or remembrance; and, if we experience change directly, then we can be realists about change.

However, a workable story along these lines has proven elusive. For example, the early-twentieth-century logician and philosopher Bertrand Russell was a supporter of this idea. He recognized the question-begging way early empiricists like Locke had accounted for the experience of time. If the concept of change comes originally from experience, then the experience of change can't itself depend on any judgment about what has happened in the past; this would presuppose an existing grasp of time and change, and so would beg the question as to how we originally acquired the idea of time and change. That is why Locke's explanation can't be right. There is a difference between experiencing change and merely judging that something has changed. We have experiences that seem like direct experiences of motion or other changes; philosopher Sean Kelly calls this the phenomenon of "pace perceived." Seeing a second hand move is not the same thing as seeing it occupy, or remembering it occupying, different locations at different times (see figure 2.3). The question is, because change is something that happens over time, how can we account for the experience of change at any moment? Are such experiences genuine, in that they are actually veridical reflections of something that is really going on? In defense of an empiricist theory of time-awareness, Russell explained the perception of motion or change as the result of our perceiving, at any moment, sense data received over a short, but extended, period of time. His view was that a sense organ, when stimulated, "goes on vibrating, like a piano string, for a while after the stimulation." In the period during which the sensation fades, we literally perceive now what happened a short while ago. It is because of this effect that we see, for example, the movement of a second hand: We can see a second hand moving, he claims, because we literally see it, at one moment, in several places.

Figure 2.3. Seeing a second hand on a clock occupying several places is not the same as seeing it move.

Russell's contemporary, H. J. Paton, pointed out that what Russell describes wouldn't actually amount to the experience of movement or change:

> [I]f in a moment I can sense several different positions of a second hand, then these different positions would be sensed as being all at the same moment. That is to say what I should sense would not be a movement, but a stationary fan covering a certain area, and perhaps getting gradually brighter towards one end. Anything else would surely be a miracle. You can't see a sensum that isn't there. If you see it, it is there at the time you see it.

Paton makes an excellent point. Getting some information now that was generated in the past doesn't give you the experience of a sequence of events—unless you are able to make an *inference* about what you have experienced that involves assigning some of that data to the past. As we have seen, any such inference presumes a grasp of the difference between present and past, thereby begging the question as to how we come to have such concepts in the first place.

The notion of an extended **phenomenological present** of experience—coupled with the failure of Russell and others—has inspired philosopher Barry Dainton to support a slightly different understanding of temporal experience in his 2000 book *Stream of Consciousness*. He rejects the Aristotelian/Augustinian notion of an infinitesimal present: However mathematically or metaphysically correct it might be to insist that any extended magnitude (such as a temporal magnitude) is in principle infinitely divisible, he argues, the notion of an infinitesimal present has no experiential relevance. A single, infinitesimal slice of information couldn't mean anything to a perceiver. Neither could one distinguish an infinite number of bits of information in any stretch of experience. Further, Dainton cites the ample evidence from empirical psychology that sequences of events that happen close enough together are perceived as simultaneous. Shorter sequences in experience, even if they are decomposable into distinct events in principle, just don't matter from the standpoint of describing what we actually perceive—of what is 'present' to us.* If we are trying to figure out how some concepts are formed on the basis of experience, we should be focusing on the nature of our actual experience, not on what is merely mathematically describable.

Edmund Husserl and C. D. Broad popularized the notion of an extended present of awareness vis-à-vis the *content* of awareness. Dainton's proposal involves extending this point to acts of awareness

---

* Remember that we are talking here about the *registration* of change, rather than its representation. One can *represent* the passage of an hour instantly by, say, simply using the phrase "one hour"; the question here is how we get a sense of the passage of time in the first place.

themselves, not just those that the act of awareness encompasses in terms of content. An infinitesimal awareness of an infinitesimal event can mean nothing to us in terms of actual experience. As Niko Strobach says, "[I]t is just as impossible to see anything at an instant as it is impossible to take a picture by opening the shutter of the lens for zero seconds." Experience itself, Dainton proposes, is constituted by overlapping, very brief, but temporally extended, acts of awareness, each of which encompasses a temporally extended streeeeetch of perceived events. This conception of experience seemingly would allow for the direct experience of change. If an act of awareness is itself extended, then a succession of experience contents can be directly apprehended; thus no need for inference or the application of presumed temporal concepts. Acts of awareness overlap in the sense of sharing common parts, so continuity of experience is preserved, as shown in Dainton's own schematic (see figure 2.4).*

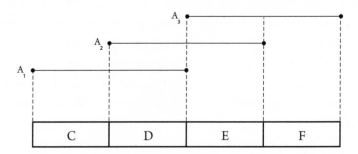

Figure 2.4

Sensory perceptions, in his view, are capable of presenting an inherent dynamism as part of the fundamental experience itself: The change, say, from D to E is encompassed by a single act of awareness

* From Barry Dainton, "Temporal Consciousness," *The Stanford Encyclopedia of Philosophy* (Fall 2010 edition), Edward N. Zalta (ed.), http://plato.stanford.edu/archives/fall2010/ entries/consciousness-temporal/. Used by permission of Barry Dainton and *The Stanford Encyclopedia of Philosophy*.

A_2. This result, if ultimately defensible, might validate Locke's project of finding the origin of the idea of change in experience.

In a related vein, Dainton's fellow British philosopher, Robin Le Poidevin, argues for a specific *sense* of motion, or change. He thinks the conjunction of a very recent memory with a present perception "gives rise to an experience of 'pure succession.' " Le Poidevin draws on the idea that there is no specific moment when information becomes conscious. Rather, different interpretations of what we perceive, at every moment, battle it out for a short while at what we might call a subconscious level, and a 'winning' interpretation eventually emerges. The winning interpretation is influenced by conditioned responses, ingrained anticipations, and additional information received later in the interpretive process. In other words, our brains take information accumulated over a short period of time and combine it into what we experience as a simple and direct consciousness of change.

One drawback of such theories is that they are impossible to prove, at least as of now. Dainton admits that we really have to wait for neuroscience to catch up in order to confirm his theory over Kant's, or over other competitors in cognitive psychology. This will probably be a long wait, given how much is yet to be understood about the brain.

As far as the big picture is concerned, however, the key point about these theories is that they are, at least, *consistent* with temporal realism. This sort of account would explain the origin of the idea of temporal succession without begging the question, as in Locke, or adverting to idealism, as in Kant. This would not itself prove that time or change are real; but if we find this reasoning plausible, it frees us to look into the possibility that there is more to time and change than temporal idealists like Parmenides, Augustine, or Kant would allow.

THE CONSTRUCTION OF TEMPORAL EXPERIENCE

These theories about the perception of change show that it is not necessarily futile to suppose that our basic temporal concepts are derived from experience, and thus that time and change are real. Even as we contemplate rejecting idealism about time, however, it is also interesting to note that we can empirically demonstrate several respects in which our experience of the order and duration of events is a mere construction on the part of our perceptual apparatus. At least in some contexts, our interpretation of time order appears to be just a kind of story told by our perceptual processing mechanisms, rather than a straightforward reflection of the order of events. At the minimum, this means that any non-idealist story we want to supply going forward must be tempered by these considerations.

There are a number of experimentally confirmed contexts in which it appears that something that happens *earlier* in one's perceptual experience is influenced by something that happens a split second *later*. In other words, there can be a demonstrable difference between what is consciously perceived and the actual temporal order of raw sensory inputs. These phenomena reveal a much more complicated story of temporal processing than we might otherwise have suspected. Here are some examples:

- **The Phi Phenomenon, aka Apparent Motion**: Two alternately flashing dots near each other typically produce the illusion of a single dot moving back and forth. (Think also of marquee lights: If they move fast enough, they produce the illusion of continuous motion.) This illusion of movement is actually very odd, as one seems to see the movement *before* the second flash, even though before the second flash there is no reason for the illusion to occur. Further, when one dot is red and the other green, the observer can get the impression that the dot changes color in midstream. The (illusory) midstream color change seems to take place, in each case, before the second flash. Seemingly even before it is perceived to occur, the second flash is somehow influencing how the sequence is experienced. (You can easily find animations demonstrating this phenomenon online, such as at http://www.philosophy.uncc.edu/faculty/phi/Phi_Color2.html.)

- **The Flash-lag Effect**: An intermittently flashing color filling a moving circle is seen as a mere crescent, while the circle is seen in its entirety. The flash, in other words, seems to lag behind the circle containing it. According to one interpretation, one's perceptual apparatus anticipates the trajectory of moving objects and registers them as ahead of their actual position; because it is stationary, the flash does not get the same treatment. (This is another effect you can find animations for online, such as at http://www.michaelbach.de/ot/mot_flashlag1/index.html.)

- **The Cutaneous Rabbit**: A subject closes her eyes and is mechanically tapped five times at a point on the wrist, then five times near the elbow, and then five times farther up the arm. (The tapping needs to be regularly timed throughout, so that there is no interruption in the pace of the tapping as it changes location on the arm.) Instead of feeling the taps at those three locations, subjects typically report feeling fifteen taps more or less equally spaced, running up the whole arm. Of course, if tapped five times at only the wrist, with no further taps, the subject reports just the five taps at the same spot on the wrist. As with the phi phenomenon, later stimuli seem to affect how the earlier stimuli are experienced.

- **Cross-saccadic Perceptual Continuity**: If you have a watch with a ticking second hand, you may experience a recurrent, momentary illusion that your watch has stopped. You look down at it, and the second hand seems to have stopped; then, after what seems like just a bit too much time, it starts ticking again. This barely noticeable yet common phenomenon has a really interesting explanation. A "saccade" is the term for the flick of one's eyes from one visual target to another. This motion, taking perhaps a tenth of a second or so, is something we do thousands of times a day. Even though our eyes are open during a saccade, we are not really aware of the visual information available to us during the movement; if we were, it would be very disorienting. The world would seem to shift in place with dizzying speed, over and over again. What the brain does (as Yarrow et al. explain in a 2001 issue of *Nature*) instead is to effectively extend the perception of the target of the saccade "*backward in time* to just before the onset of the saccade" (emphasis added). What happens is that the perceived duration of the event you are now looking at gets extended by about the same amount of time it took to move your eyes to it. The intervening information never makes it to consciousness, thus preserving a temporal continuity between pre-saccade and post-saccade perceptual consciousness (see figure 2.5). Usually, we don't notice this effect. But where there is an external time reference like a ticking second hand, the artificial extension of the second hand's perceived movement can sometimes make it seem as though the second hand takes more than a second to tick forward, thus the momentary illusion of a stopped watch.

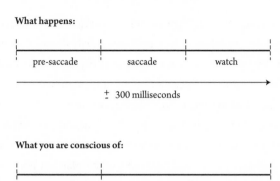

What happens:

pre-saccade   saccade   watch

± 300 milliseconds

What you are conscious of:

pre-saccade   watch

± 300 milliseconds

Figure 2.5

We need to be careful in describing just what is happening in cases like these. Is your memory playing tricks? Are intervening experiences causing you to retroactively reinterpret the perceived order of events? Is anticipation causing you to misinterpret some stimuli? Or perhaps different interpretive pathways in the brain simultaneously form multiple, inconsistent interpretations over the course of the stimulus, and one of them wins out just because it reflects the sort of interpretation that is simplest or most helpful in most situations. One can speculate on good reasons why temporal processing would have evolved to work this way. Take cross-saccadic continuity: Without it, our experience would be full of disorienting and useless visual information. In a similar vein, our need to take delays in processing into account while tracking moving objects could explain the flash-lag effect.

These surprising experiential illusions continue to receive attention, from neuroscientists and others. What are the philosophical implications?

On any account, these illusions definitely show that time order as we know it in experience does not necessarily correlate with time order in reality and can even systematically deviate from it. You would expect natural selection to generally favor accurate portrayals of reality, but evidently there are exceptions to this rule. Note that there is nothing in the phenomenological present account that would explain, even in principle, how this sort of real-time rewriting of experience is supposed to take place. It goes to show how far we have to go to give an adequate account of temporal awareness.

What we learn from these phenomena is that time order in experience is a sort of construction, which is what an idealist like Kant would like to claim. At the same time, it should be stressed that these experiments do not provide any sort of direct support for temporal idealism. The very description of these cases involves distinguishing between experienced and objective time, whereas an idealist rejects the notion of objective time altogether.

So this sort of phenomenon doesn't prove that time order and duration is itself an invention, just that particular judgments of order and duration derive from a post-hoc process involving different cognitive and neurological modules, only some of which are primarily tasked with accuracy in representation. Thus, although these experiments create a confusing complication for realist theories of time, they teach us more about the mind than about time. Though they show that our contact with the 'real' time order of events is rather less straightforward than we might have thought, they do not require us to embrace temporal idealism either.

Then what about temporal realism? All we have done so far is to show that realism about time is not obviously inconsistent with what we know about time-awareness. We have yet to discuss a positive theory about what time *is*, if indeed it is something. Thus we turn in the next chapter to early-modern and 20th-century physical theories of the universe, revisiting the question "Is time real?"

WORKS CITED IN THIS CHAPTER

Dainton, Barry. *Stream of Consciousness* (New York: Routledge, 2000).

––––. "Temporal Consciousness," *Stanford Encyclopedia of Philosophy*, ed. by Edward N. Zalta, http://plato.stanford.edu/entries/consciousness-temporal.

Kant, Immanuel. *Critique of Pure Reason* (1787).

Kelly, Sean. "The Puzzle of Temporal Experience," in *Cognition and the Brain*, ed. by Andy Brook and Kathleen Akins (Cambridge, UK: Cambridge University Press, 2005).

Le Poidevin, Robin. *The Images of Time* (Oxford, UK: Oxford University Press, 2007).

Locke, John. *An Essay Concerning Human Understanding* (1690).

Paton, H. J. "Self-Identity," *Mind* 38 (1929), 312–329.

Russell, Bertrand. *An Outline of Philosophy* (New York: Routledge, 1996).

Strobach, Niko. "Zeno's Paradoxes," in *A Companion to the Philosophy of Time*, ed. by Adrian Bardon and Heather Dyke (Oxford, UK: Wiley-Blackwell, forthcoming).

Yarrow et al., "Illusory perceptions of space and time preserve cross-saccadic perceptual continuity," *Nature* 414 (2001), 302–305.

Other works of relevance to the issues in this chapter

Broad, C. D. *An Examination of McTaggart's Philosophy, Vol. II* (Cambridge, UK: Cambridge University Press, 1938).

Dennett, Daniel. *Consciousness Explained* (Boston: Back Bay Books, 1992).

Dicker, Georges. *Kant's Theory of Knowledge: An Analytical Introduction* (Oxford, UK: Oxford University Press, 2004).

Grush, Rick. "Time and Experience," in *The Philosophy of Time*, ed. by Thomas Müller (Frankfurt, DE: Klosterman, 2007).

Husserl, Edmund. *The Phenomenology of Internal Time-Consciousness* (1917).

Nolan, Christopher. *Memento* [movie] (1999).

Paul, L. A. "Temporal Experience," in *The Future of the Philosophy of Time*, ed. by Adrian Bardon (New York: Routledge, 2011).

Chapter 3

Time and Space-Time

Thanks to its being championed by the Catholic Church, Aristotelian physics in the West persisted as pretty much the unchallenged approach to understanding nature through the Middle Ages. This situation began to change with the arrival in the seventeenth century of the "Scientific Revolution," the rebirth of experimental and mathematical science in Europe after religious authorities had quashed or redirected the pursuit of knowledge for over a thousand years. There were a number of great (and often courageous) scientists associated with this movement—such as Giordano Bruno, Galileo Galilei, Francis Bacon, and Robert Boyle—but Isaac Newton was, in many ways, the most successful of them all. Although other scientists had disproved certain elements of Aristotelian or Catholic dogma (most famously, the geocentric model of the universe), Newton produced a comprehensive system of physical laws that totally superseded what had been the received view since ancient times. This system persisted until Einstein, whose revision of Newtonianism took our understanding of time to a whole new level.

REPLACING ARISTOTLE

Newton's greatest and most synoptic achievement was his execution of the idea (earlier conceived by Galileo) that there could be mathematically describable laws of motion. These laws would be quantifiable, invariable, and universal, and would involve just mechanical and gravitational interaction. This was a huge break from traditional Aristotelian physics, which did not require universal laws and explained phenomena primarily by reference to intrinsic properties of things, rather than by mechanical interactions and forces acting according to uniform rules. As we shall see, Newton's views on the nature of time are intimately bound up with this project.

In his *Philosophiæ Naturalis Principia Mathematica*,* commonly known as the *Principia*, he identified a connection between the possibility of these universal laws of dynamics and the very real existence of time and space as entities in their own right. Most important, he was convinced that, to have the status of universality, his rules would have to pertain to what he called "absolute" and "true" motion rather than mere "relative" motion. The concept of **relative motion** is easy to understand. A sailor walking forward on a ship at one foot per second is moving one foot per second relative to the ship; if the ship is moving forward at ten feet per second relative to the ocean, then the sailor is moving eleven feet per second relative to the ocean. Relative motion is motion merely with regard to some other body, or with regard to a 'place' defined by some other body or bodies (such as a ship's quarterdeck, or the corner of Fifth and Main). Being in motion relative to some other thing doesn't necessarily mean you are truly in motion: For example, someone sitting still while being passed by a train is in motion relative to the train.

Relative motion stands in contrast to **absolute motion**. Newton describes the scientific measurement of absolute motion in terms of the only values by which such motion could be measured: **absolute space** and **absolute time**. That is to say, absolute motion is

* *Mathematical Principles of Natural Philosophy.*

motion through absolute space, with the rate of motion given in terms of absolute time. He introduces these concepts just before the statement of his three famous laws of motion (familiar to anyone who took high school physics). "Relative space" refers to the space taken up by a body at any point, or to a place defined in relation to the position of other bodies; relative motion is motion relative to some place or other body. Absolute motion of a body, by contrast, is motion with respect to "immovable space." This space is the (presumed) underlying three-dimensional field in which all bodies have some absolute position, independent of anything else's location. "Relative time," which he also calls "apparent" or "common" time, refers to the measurement of durations or processes by reference to some other motion, such as the movement of a clock's minute hand. This he distinguishes from absolute time, which "of itself, and from its own nature, flows equably without relation to anything external." Absolute time is needed, he continues, in sciences like astronomy as a corrective to common time. For example, the length of a day varies and so cannot be used for mathematically precise astronomical formulas or predictions.

What does it mean to be a realist about space and time? Newton did not think space and time were material substances, like rocks or trees. Rather, he thought of them as sui generis entities—entities of a kind all their own. They essentially function, for him, as universal containers for, respectively, bodies and events. As such, they are not themselves bodies or events, or constituted by bodies or events, but they are real and a precondition for the possibility of what they contain.

This means that, although events in time are therefore dependent for their possibility on the existence of time, time itself with its inexorable "flow" does not depend on events. This makes sense only if one accepts that it also makes sense to say that time would still pass in a universe devoid of any motion or other change. Thus the very notion of absolute time is a repudiation of an Aristotelian understanding of time; on the Aristotelian understanding, because time is just the measure of change, the notion of time passing without change is incoherent. This is, therefore, a fundamental departure from Aristotle's relationism.

Newton's rejection of the Aristotelian approach has to do with the usefulness of time as a metric in describing motion and the forces involved in motion. His reason for embracing the concepts of absolute space, time, and motion was that he thought that universal laws of motion were describable only in such terms. Laws of motion referencing mere relative motions could not apply universally, because relative motion is determined by the contingent motions of other bodies. The explanation of *changes* in motion (i.e., acceleration) by reference to forces is the heart of Newton's system. This, in turn, requires absolute motion, because relative motion doesn't require a force being applied to any particular body. Thus an immovable, real space must be presupposed. Further, because any motion, such as the rotation of the Earth, can be irregular,* the time cited in universal laws of motion cannot depend on the motion of any particular object. Hence the need, if Newton's whole project is even to be possible, to presuppose a metric—absolute time—independent of the motion of any object. "True" time, meaning an actual, independent entity, provides an objective basis for this universal metric for measuring motion.

On Aristotle's relationist view, time is a mere abstract measure of change. Newton's theory, by contrast, treats time as a real thing unto

---

* Note that any motion can seem regular if it is being measured with a device that is irregular yet synchronized with the motion it is measuring. For example, you wouldn't be able to tell that the Earth takes a second longer to rotate every day if all you have to measure the period of its rotation is a clock that loses a second every day.

itself. With this theory, Newton has fundamentally broken with the Aristotelian perspective holding time to be dependent on change; change is now best described as something that happens *in* time, even as time itself flows along. For an idealist like Kant, the most accurate way of talking about time would be adverbially, as a way of describing how we experience things. Newton treats time like a *thing* itself—not, indeed, a material substance like a tree, a cow, or a body of liquid, but still something that has noun status. His understanding of time is therefore radically different from either a relationist or idealist conception: This is the return of temporal **realism.**

So far, we have only described the concept of absolute time as a presupposition underlying (in Newton's mind, at least) the possibility of universal laws of dynamics. What reasoning or evidence did he present in support of the claim that absolute and true time really exists?

Most fundamentally, he thought that there must be absolute time if there is evidence for the existence of absolute *motion*. He noted the action of water in a rotating bucket: The surface of quickly rotating water becomes concave as it pushes out toward the sides of the bucket (see figure 3.1).

Newton pointed out that the spinning water will demonstrate this concavity when it is spinning just as fast as the bucket itself. Thus the water's concavity can't be explained by reference to the relative motion of the water versus the motion of the bucket. He also cites the outward force you seem to feel while swinging a weight on a cord in a circle around you, aka the "centrifugal effort." Where is that force coming from if the weight is not really moving (i.e., in absolute terms)?

All Newton could directly conclude from phenomena like these was that rotational motion cannot be explained just by reference to the rotating object's motion relative to its immediate material

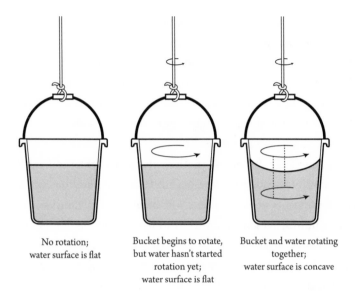

No rotation;
water surface is flat

Bucket begins to rotate,
but water hasn't started
rotation yet;
water surface is flat

Bucket and water rotating
together;
water surface is concave

Figure 3.1

surroundings. However, one interpretation of these rotational effects is that the water is experiencing true, absolute motion—defined as motion relative to absolute, immovable space itself. This is the conclusion Newton apparently favored. And, if there really is absolute motion in Newton's sense, then every such motion must take some fixed amount of time. Further, this time measure—because it correlates with absolute motion and describes all true motion—must be distinguished from any time measure derived from any particular regular(-ish) motion we observe, such as the orbit of the Moon or the movement of the hands of a clock.

In fact, Newton thought he saw concrete evidence of an absolute, underlying regularity to the passage of time in the synchrony of certain kinds of motions. We have said that the solar day is irregular, but irregular compared to what? It is irregular compared to various astronomical motions, such as the orbits of the moons of Jupiter, or to a pendulum clock. But why not suppose that the solar day is regular and these other processes are the irregular ones? It is simply because these other processes agree so nicely with each other. Newton thought that the best explanation for the agreement between certain kinds of otherwise unconnected motions was that they were somehow tracking the absolute, regular flow of universal time.

Absolute space and time allow real (non-relative) motion to be explained by consistent rules with mathematical precision. Further, real, absolute space and time permit laws of nature that are actually descriptive of nature; in contrast, Aristotle's time is a mere abstraction by which we understand or measure natural phenomena. Thanks in large part to Newton's work, Aristotelian physical science, which had dominated Western science for nearly two thousand years, was supplanted by a system of rules of nature that were precise, explanatory, and yielded testable predictions; this is why Newton is the most important figure in the history of modern science—*and we now understand that his entire project stands or falls with his conception of time.*

NEWTON AND LEIBNIZ

One of Newton's contemporaries was the remarkable Austrian logician, theologian, and all-around polymath Gottfried Wilhelm Leibniz, who had his own, very different ideas about physics and criticized Newton's conclusions from afar. By 1715, these two renowned thinkers had been part of a prolonged, nasty, all-out international conflict over who should have credit for the development of calculus (Leibniz had published his version first, but Newton had independently developed a slightly different method for solving the same kinds of problem years before). He and his intermediaries had accused Leibniz of plagiarism. Further, Leibniz was a vocal critic of Newton's physical theories, and for this he was attacked by Newton's many supporters.

Fortunately, Princess Caroline—the Princess of Wales at the time, and a student of philosophy herself—was friendly with both Leibniz and a number of Newtonians. She was able to arrange a public correspondence between Leibniz and Samuel Clarke, a friend of Newton who was understood to be speaking for him in the correspondence. The subject of the letters, published in 1717, was Clarke's defense of Newton's doctrines of time and space against Leibniz's theological and philosophical criticisms. In his contributions, Leibniz defends an Aristotelian conception of time, though, unlike Aristotle, his motivation for rejecting realism about time was primarily theological in nature.

Leibniz's main objections to realism about time could be summarized as follows:

- Time cannot exist in itself because, were time real, it would still only exist at any moment in the form of an instant; and nothing can be composed of instants (as Aristotle had explained in response to Zeno). If time can't be composed of instants, then what is it made of? Not matter or energy, or "immaterial spirits" (as Clarke proposes at one point); but what else is there?

- Newton had referred to time and space rather vaguely as "emanations" of God. Leibniz argues that, if time were a thing in itself, and God is 'in' time, then God would be dependent for His existence on the existence of time. And God, being perfect, must be fully self-sufficient.

- If time were an independent existent, then there would be a question as to why God created the universe at one time rather than another. God being perfect, He could not act without reason—that is, whimsically. But (as Augustine had also noted) if time exists for God, then there could be no reason why He would choose one moment of empty time to get things started, rather than another. So this conception of time must be wrong.

Leibniz adds some parallel points about the idea of absolute space. Absolute space would have to be composed of points; God would have to be in space, and so dependent on it; absolute space means God would have needed to make the arbitrary choice as to the orientation and location of the entire contents of the universe as a whole—Which way, east or west? Place everything where it is now, or three feet over to the right?

Leibniz's alternative is a conception of time and space as a set of relations holding between, respectively, events and objects. He was not a realist, in that he denies that time exists in itself. Neither is he an idealist, in that, like Aristotle, he thinks the time relation is a legitimate category for scientific and mathematical accounts of the universe. Thus we classify Leibniz as a relationist, just like Aristotle; the major difference between the two is that, like Augustine's, Leibniz's conviction about the nature of time is based primarily on a perceived conflict between temporal realism and religious dogma.

In response to Leibniz's objections, Clarke points out that, in both everyday and scientific parlance, time is treated as a "quantity" (i.e., something that can be measured and that there might be more or less of). Leibniz, he claims, also ignores the empirical evidence (such as rotational effects) for absolute motion, which in turn implies absolute time.

Leibniz hated the idea of absolute motion. Imagine a universe of bodies all in (absolute) motion in the same direction and at the same rate—imagine, say, that everything in the universe is drifting to the left at five meters per hour. There would be no effective difference, Leibniz insists, between this scenario and one in which everything is at absolute rest. The fact that the concept of absolute motion appears to make these scenarios meaningful shows it to be incoherent, he claims, for two reasons: First, the difference between the two situations has no observable consequences, and second, God would have no reason to prefer one over the other, and so would necessarily act without reason if forced to pick one over the other.

However, as Clarke repeatedly points out, Leibniz has no convincing explanation for the water's centrifugal effort in Newton's spinning bucket. All Leibniz can say is that, well, such is the nature of circular motion: It exhibits a force that other sorts of motions do not. Newton's laws, in conjunction with his postulation of a universal gravitational force, do a very good job of explaining and predicting observable phenomena. Leibniz had his own theories of motion; we won't go into them here, but it suffices to say that they had much less explanatory and predictive power.

Newton believes in time because of its usefulness as a metric. But what does he think it is? Even in his absolutist language, time is defined more in terms of what it *does* (i.e., "flow," or pass, or lapse) than in terms of what it *is*. This is pretty typical when it comes to discussions of the nature of time. Partly for this reason, the question about space tends to be "Is space real?"; whereas the question about time (instead of "Is time real?") often becomes "Does time really pass?" But the latter presupposes the main question. If we want to coherently think of time as a thing that *passes*, it would seem we need to think of it as a *thing*, at least in some sense of 'thing.' As difficult as it is to handle the question "Is space real?" it is even harder with time. In a way, it is not too hard to vaguely imagine space as a medium—as a kind of insubstantial soup in which we float around like croutons. What is the analogous image for time that allows us to (even vaguely) represent it as a real item in the world? Lacking any way to get a grip on the idea of time as a thing, how do we reconcile Newton's success at a comprehensive theory of motion and gravity with his failure to offer a theory of the nature of time, when his physics is so intimately bound up with the existence of absolute time?

Post-Newtonian physics may help with the problem by replacing Newton's absolute space and time with the concept of relativistic **space-time**. To understand the modern concept of space-time, and how it might help us decide if there really is such a thing as time, we need to look at how Newtonian physics was superseded by Einstein's theory of relativity.

RELATIVITY

Newton's theory remained the almost universally accepted account of motion and gravity until the late nineteenth century, when it began to be questioned by experimental physicists. The key discovery that led to Albert Einstein's revolutionary theory had to do with light. Einstein's breakthrough had to do with the combination of a couple of facts about light plus one extraordinary experimental finding about its behavior. First of all, as Newton was aware, the speed of light is finite; the Danish astronomer Ole Christensen Roemer had identified and reported this fact in 1676. Second, by the late nineteenth century, it had become widely accepted that light was a variety of electromagnetic radiation, and that it propagated itself in a wavelike manner. The experimental finding that then opened the door to Einstein's theory was that the speed of light is a constant regardless of the velocity of the source.

Some reasoned that, if light moves in a wavelike manner at a finite speed, there must be some medium through which it travels; if ocean waves, for example, are disturbances of water, and light is like a wave, then light must be a disturbance of something. The proposal considered by American physicists Albert Michelson and Edward Morley in 1887 was that light (or any electromagnetic radiation) propagates itself through an underlying, invisible, and fixed medium called the **aether**. (Some proposed that the aether was simply Newton's absolute space.) If so, one would expect that light measured in the direction of its motion should appear to travel more slowly, and light coming from a source toward which we are moving should appear to be going by faster—much as ripples in a pond will appear to move more slowly if we are moving along with them, and will appear to go by more quickly (in the opposite direction) if we are moving toward their source. Michelson and Morley were amazed to discover that the speed of light is constant: In a vacuum, *light travels at the same speed regardless of the velocity of the source of the light or one's velocity relative to the source of the light.* If, say, you throw a ball forward at ten miles per hour while running in the same direction at ten miles per hour, the ball will be traveling at twenty miles per hour. But light travels away from its source at a constant speed regardless of what the source is doing. Unlike an object launched from a rapidly moving vehicle in the direction of travel, a light beam projected from that vehicle will travel at the same speed as if it were projected from a resting source.

This result tends to undermine the notion that the propagation of light has anything to do with an aether. You can't overstate the importance of this discovery, which physicists initially resisted. Newton had accepted that, from any observer's viewpoint, *mechanical* laws of nature will always operate the same way, so there is no method involving the behavior of *objects* that determines absolute (vs. merely relative) constant velocity. The Michelson-Morley result showed that differences in the speed of light couldn't be used to establish absolute velocity either, and so the notion of an underlying, fixed aether (or a Newtonian absolute space) was inoperable in describing or predicting motion.

This is where Einstein comes in. While he was still working as a patent clerk in Switzerland, he worked out the basics of a new theory about the nature of space and time. It starts with a theory about motion. Where others tried to find some loophole, Einstein took seriously what the Michelson-Morley experiment had suggested: that the speed of light in a vacuum is constant regardless of the velocity of the source. He proposed that the notion of absolute velocity, being irrelevant, should be dispensed with altogether.* Without an aether, or some other stand-in for absolute space, absolute motion is not only undetectable but also meaningless. Einstein proposed that all laws of nature (meaning all mechanical laws, plus the newly discovered rules governing electromagnetic phenomena like light, plus anything else) are the same for any observer regardless of his/her constant velocity relative to anything else. This simple proposal— since confirmed many times over—has profound consequences for our understanding of time and reality.

Once you examine the consequences, it becomes evident that the extraordinary fact that the speed of light in a vacuum is a constant, when added to the elimination of the idea of absolute motion, pretty conclusively rules out any notion of a Newtonian absolute time. This conclusion is already evident in Einstein's limited **special**

---

* This refers to constant velocity, not acceleration. Thus it does not directly apply to circular motion (which is accelerated motion) as in Newton's bucket. Newton failed to note that the water in the bucket is rotating relative to the rest of the universe; some have argued that this relative motion could explain centrifugal effort in such cases. If so, then the effects of rotational motion don't prove absolute motion the way Newton thought they did.

theory of relativity, which amounts to a simplified account of time and space that doesn't concern itself with mass and gravity (as the **general theory** does). To use a familiar illustration, imagine two observers moving relative to each other: One (Albert) is on a train traveling west to east, and the other (Isaac) sits on the train platform. The following describes Isaac's experience only. Just as the train passes, two bolts of lightning strike: one a mile to the west and one a mile to the east (see figure 3.2).*

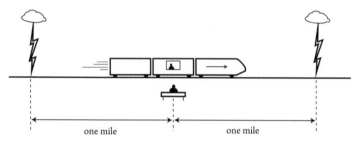

Figure 3.2

* These distances are as Isaac measures them; as we shall see, distances, like times, are relative to one's frame of reference.

A very short time later, Isaac sees both flashes simultaneously. Knowing the distance to each strike, he concludes, reasonably enough, that the two strikes occurred simultaneously. But is he right? Albert disagrees with Isaac; for Albert, the flashes do not arrive simultaneously. Isaac thinks he has a simple explanation for Albert's confusion: The light coming from the west had to catch up with Albert, and Albert is approaching the light coming from the east (see figure 3.3).

This answer would work just fine, *if* you could say that Isaac is really at rest and Albert is really in motion. But we cannot, according to Einstein's proposal. Remember that there is no aether, and no way to determine absolute velocity; Einstein's hypothesis is that this is because *there is no such thing*. Albert can equally well say that

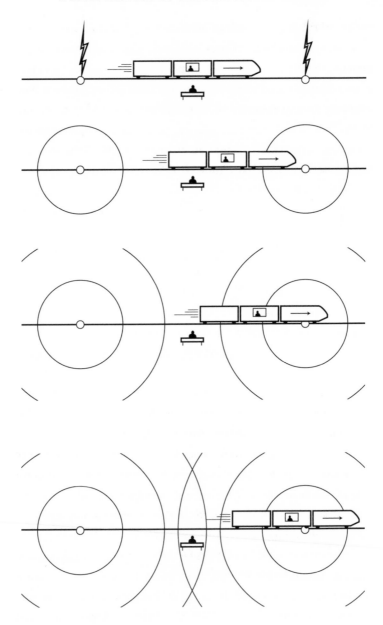

Figure 3.3. Isaac's diagnosis of Albert's 'mistake.'

he is at rest and Isaac is in motion as the Earth flies by beneath the train. Because the equidistant lightning strikes arrived at Albert's location at different times, Albert is as justified in saying they were *not* simultaneous as Isaac is in saying that they *were* simultaneous. Without absolute motion, neither is right and neither is wrong. They just have different frames of reference; each describes the situation from the standpoint of his frame of reference, and there is no authoritative, independent standpoint that would decide the matter. This means that there is no absolute, objective fact of the matter as to whether or not the flashes are simultaneous: Simultaneity is relative.

If simultaneity is relative, then there is no absolute time. No one can say when a given event occurs, or what time it is now, with an authority that transcends one's frame of reference. *What time it is 'now' depends on one's inertial frame of reference.* Another well-known thought experiment illustrates this fact. Imagine a simple clock that measures time with a light pulse reflected back and forth inside a vertical tube (figure 3.4).

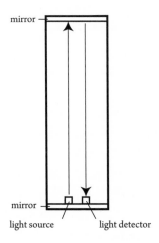

mirror

mirror

light source   light detector

Figure 3.4

Figure 3.5

Now, put a clock like this on the train with Albert, and compare Albert's experience of the clock with Isaac's as the train passes by (figure 3.5).

Note the path of the light pulse from Albert's perspective versus its path from Isaac's perspective (figure 3.6).

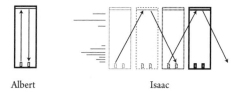

Albert   Isaac

Figure 3.6

From Isaac's perspective, the light beam in Albert's clock has farther to travel for each round trip than the one in his clock. Consequently, Isaac sees Albert's clock as ticking more slowly than his own; according to Isaac, Albert's clock is slow. But suppose Isaac has a light clock just like Albert's. From Albert's perspective, the path of the light pulse in Isaac's clock looks just like the path of the light in Albert's clock as seen by Isaac. For Albert, Isaac's light clock is the slow one. In fact, they would each be wrong in finding the other's clock to be slow, because in either case their finding presumes absolute motion and absolute time. Neither Albert nor Isaac can be said to be in a state of either absolute motion or rest. They simply have different frames of reference. But this means that the rate of time's passage is relative to one's frame of reference. Once again, there is no single, objective time.* There is no definitive 'God's eye' view that would, even in principle, tell you who is right and who is wrong as to your absolute velocity or what time it is.

It is important to be clear on what is happening here. Going fast doesn't make time slow down. This would be mistaken in two respects. First of all, there is no such thing as going fast in absolute terms: Going fast versus being at rest are relative. Second, there is no absolute time to be sped up or slowed down; there is only my measurement and your measurement, which simply will not agree if we have different relative velocities. This does not mean that the time of events is entirely subjective. Two observers in the same reference frame (and near each other in space) should agree as to what time it is and which events are simultaneous with each other.

Einstein made a lot of very specific predictions on the basis of his supposition about the relativity of motion, and observations on the level of large-scale events have confirmed his theory again and again. His theory dispenses with any need for an objective background flow of time and provides no basis for true or false claims about what time it is independent of one's frame of reference. In

---

* There is an easy answer as to why this conclusion seems so awfully counterintuitive. Relativistic effects like the ones we are discussing here would only be observable given extremely high relative velocities and/or extremely sensitive instruments. We didn't evolve under conditions where these effects were either noticeable or relevant, so we wouldn't expect our perceptual or cognitive faculties to be adapted to a world in which these effects matter. As a result, it feels distinctly unnatural when we try to tackle relativity either imaginatively or intellectually. Yet relativity has been experimentally confirmed over and over again.

modern, relativistic physics, space and time are replaced by **space-time**. The main reason for this is that spatial and temporal quantities individually vary according to one's frame of reference; there will, however, be an invariant quantity (known as a **space-time interval**) that can be decomposed, or sliced, into spatial and temporal intervals in innumerable ways according, again, to one's frame of reference. The concept of space-time is useful for this reason: Although different observers will carve up space and time in different ways, in that their paths through space and time will depend on their perspectives, all can agree on paths described mathematically in terms of four-dimensional space-time. Furthermore, although velocity in space is relative, velocity in space-time is not. In a somewhat loose manner of speaking, whereas nothing can go faster through space than light, everyone and everything is 'moving' at the speed of light through space-time. Although how much of one's constant velocity is velocity through time versus velocity through space depends on one's frame of reference, the combination of one's motion through space and one's motion through time always adds up to the speed of light.

Relativity thus tells us we must think in terms of a real quantity, space-time, in order to address the ancient question of the reality and nature of time. Or, to put it a little more modestly, the stipulation that there is such a thing as space-time works in creating an effective mathematical model of what is going on. Now we need to understand space-time a little better, because we need to get into position to ask the question: Is space-time real?

WHAT IS SPACE-TIME?

In support of Einstein's theory, Polish mathematician Hermann Minkowski worked out the mathematics of space-time around 1908. In a lecture he said the following:

The views of space and time that I wish to lay before you have sprung from the soil of experimental physics, and therein lies their strength. They are radical. Henceforth space by itself, and time by itself, are doomed to fade away into mere shadows, and only a kind of union of the two will preserve an independent reality.

We have seen that observers in different inertial frames, like Isaac and Albert, will disagree on the temporal intervals between events. They will also disagree about spatial intervals. Because the speed of light is constant for any inertial frame, spatial distance can be measured by the amount of time it takes light to travel from one point to another; thus the concept of a 'light-year' (i.e., the distance traveled by a beam of light in one year). This means that there is a deep connection between the measurement of temporal intervals and the measurement of spatial intervals. If Isaac and Albert can't agree on what time it is, then they will also have different readings on spatial distance. They will not agree on the elapsing of one year, and so what Isaac measures as a light-year, or fraction thereof, will not be the same as what Albert measures as one.

All is not chaos in determining spatial or temporal intervals, however. *Spatiotemporal* intervals are constant for all observers: An observer in one inertial reference frame might measure a smaller spatial interval, but a larger temporal interval between two events, and vice-versa for another. So, spatial and temporal intervals vary in an orderly way to allow for a unified and universal space-time interval between all events—one that all observers can agree on. This is what Minkowski meant when he said that only a union of space and time "will preserve an independent reality."

It would be nice, at this point, to be able to draw a diagram showing how objects and events are arrayed in space-time. However, although space-time can be represented mathematically, it is not possible to produce an adequate pictorial representation, because each observer sees it differently. One could try picturing a loaf of bread carved at different angles, but that would really represent a case of carving up three-dimensional space differently, not four-dimensional space-time. Minkowski came up with a clever way of visually representing space-time, though only from the perspective of a particular observer: the **light cone**. From a given perspective, all of reality can be divided into three components: one's past light cone, one's future light cone, and one's **absolute elsewhere**. Let's take Isaac again. What falls into Isaac's 'absolute past' (i.e., anything within his past light cone) is limited by the speed of light, which relativity predicts cannot be exceeded: The past light cone indicates the spatial distance of those events that could have affected Isaac by now or of which he could be aware.* Imagine beams of light arriving at Isaac's location from all directions; the surface of the past light cone represents the path of those beams, and the inside of the past light cone represents all slower-than-light events capable of affecting him. Conversely for his future light cone: The surface represents the spatial distance of future events that he could influence by sending out a signal at the speed of light.

Figure 3.7 represents space-time from the perspective of any generic observer, such as Isaac and/or anyone else at his location and in his inertial frame. (Note that the past and future light cones should be understood three-dimensionally: Isaac's future light cone, for example, really represents an expanding sphere of possible influence.) We say that the events within Isaac's past and future light cones are **time-like separated** from Isaac. The absolute elsewhere—that

---

* There is no 'correct' inertial frame, so, despite the common use of the terms 'absolute past' and 'absolute future' in the context of Minkowski space-time, there is no absolute past or future in the *Newtonian* sense. In this context, these terms just mean that any observer will agree that certain events lie in *Isaac's* past and future, respectively.

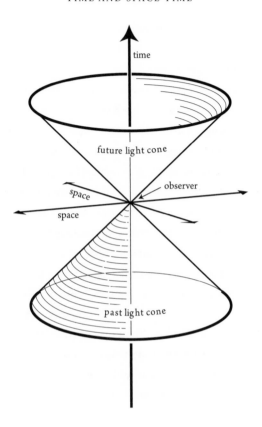

Figure 3.7

area that falls outside Isaac's light cone—represents the spatial/ temporal area such that people at his location, but moving at different speeds, will disagree on which event in that area is simultaneous with a given event occurring at his location; we say that events falling outside of his light cone are **space-like separated** from him. (The lightning strikes in the earlier illustration, for example, were space-like separated from both Albert and Isaac.) In this way, we can place everything into one space-time and explain why different perspectives on things like velocity differ by reference to how each observer slices up space-time into what is time-like and what is space-like separated from him or her.

The concept of space-time as a real venue for things and events becomes even more important in the context of Einstein's expanded general theory of relativity, which offers a new theory of gravity and, in so doing, accounts for the subjective effects of mass, gravity, and acceleration. Unlike constant velocity, acceleration is not a merely frame-dependent issue, and so its objective characteristics need to be differentiated from velocity. The way Einstein does this is in terms of trajectories though space-time: Unlike constant velocity, acceleration is expressed in terms of a nonlinear trajectory though space-time. Further, in the general theory, gravitation is explained in terms of distortions in space-time related to the presence of mass. Space-time is thus a very useful tool, mathematically speaking, to explain both the relativity of velocity and the non-relative forces of acceleration and gravity. In contemporary physics, it is common to talk in terms of a real space-time that contains, or constitutes, the universe of objects and events as we know it.

SPACE-TIME REALISM

We will come back to the theory of relativity in the next chapter when we talk about the reality of the passage of time; the implications of the relativity of simultaneity will come into play once again there. In the meantime, let us focus on the idea of space-time itself. Now that we have dispensed with Newton's absolute space and time, what about Einstein/Minkowski space-time? Is it a real thing? Physicists typically do not focus heavily on this question, because the success of their models in explaining and predicting phenomena doesn't require an answer to it. Yet some physicists, and many philosophers, have tried to tackle this issue as best they can. We have seen that relativity treats space-time as a sort of four-dimensional block—with properties of its own—that is carved up differently into its spatial and temporal components, depending on one's frame of reference. *But if it is a block, what is the block made of?* Answers to this question tend to get rather abstruse. One classical realist proposal is that space-time is constituted by momentary space-time points; a variation on this proposal is that space-time is made up of space-time points but also involves the metrical relationship between those points as part of its identity. (The latter version is intended to solve a major problem with the simpler version, in that the simpler version appears to leave the actual distribution of points in space-time inert for explanatory or predictive purposes.)

The replacement of Newton's absolute space and time with space-time would not satisfy Leibniz: He would probably want to ask whether it would be possible for the entire universe of material bodies to be uniformly oriented differently with regard to this real space-time. The question of whether we inhabit this universe or its mirror image has no practical or observable consequences. Yet, anyone who is a realist about space-time has to take this sort of possibility seriously. Leibniz would complain about such a possibility on theological grounds (pertaining to God not having any reason to choose one distribution over another); but one could also complain, from a scientific standpoint, about being told there were two different physical possibilities (i.e., this universe or its mirror image) with no possible observational difference. Philosophers of science commonly express doubt that any purported scientific theory that allows such alternate possibilities can be considered a satisfactory physical theory.

Further, recall that Aristotle had pointed out the impossibility of anything being composed—literally constructed out of—infinitesimal points. A Euclidean point has zero length, and no number of items of zero length add up to something of non-zero length. Composition out of points gives us an infinite number of definable, finite spatial segments between any two points, and thus gives us Zeno's paradoxes of "The Dichotomy" and "Achilles and the Tortoise." This would seem to rule out any view of space-time as composed of points. How can space-time be composed of points if a collection of points cannot add up to anything? If the bits making up space-time are not mere mathematical fictions like points, then what are they?

In light of these concerns, it seems unlikely that we can meaningfully think of space-time as composed of units the way a brick wall is composed of bricks. Barry Dainton muses that "it may be that not all real substances can be broken into parts in the way of macroscopic material things." In other words, if one wanted to think of space-time as a real thing, one might need to think of it as something that doesn't have parts the way a house—or even a hydrogen atom—does. Is this a coherent possibility? If something is a thing, must it not have parts, at least in principle? Or does this rule only apply to things *in* space-time, and not to space-time itself?

In considering space-time realism, perhaps the most fundamental question is whether capturing reality is actually the point. We have seen that a typical interpretation of relativity treats space-time as an entity in its own right. **Scientific realism** is the view that successful scientific theories are to be taken literally: The entities that they describe, both the observable (like comets or three-toed sloths) and unobservable (like electrons or space-time), are to be taken to truly exist as described.

The alternative is that the point of fundamental physical science is just to create a model that is essentially pragmatic, in that its purpose is just to help us systematize observations and predict events as we know them (and only as we know them). This alternative to scientific realism is called **scientific instrumentalism**. One might have assumed that scientists themselves would usually be scientific realists, but in fact this is often not the case. In a recent essay, physicist N. David Mermin takes an instrumentalist position:

> The raw material of our experience consists of events. Events, by virtue of being directly accessible to our experience, have an unavoidably classical character. Space and time and space-time are not properties of the world we live in but concepts we have invented to help us organize classical events. Notions like dimension or interval, or curvature or geodesics, are properties not of the world we live in but of the abstract geometric constructions we have invented to help us organize events.

Whether we take physicists' pronouncements about the nature of space-time to be relevant to reality depends on whether we take reality to be the subject matter of physics. The right answer to this will likely be nuanced and may require a modest interpretation of physical theories. In *A Brief History of Time*, Stephen Hawking offers a modest description of a scientific theory:

> [A] theory is just a model of the universe, or a restricted part of it, and a set of rules that relate quantities in the model to observations that we make. It exists only in our minds, and does not have any other reality (whatever that might mean). A theory is a good theory if it satisfies two requirements: it must accurately describe a large class of observations on the basis of a model that contains only a few arbitrary elements, and it must make definite predictions about the results of future observations.

In *From Eternity to Here*, Sean Carroll's 2010 book on the physics of time, he agrees, and references the implications for temporal concepts:

> Perhaps surprisingly, physicists are not overly concerned with adjudicating which concepts are 'real' or not. They care very much about how the real world works, but to them it's a matter of constructing theoretical models and comparing them with empirical data. It's not the individual concepts characteristic of each model ('past,' 'future,' 'time') that matter; it's the structure as a whole. Indeed it often turns out to be the case that one specific model can be described in two completely different ways, using an entirely different set of concepts.

These statements by Mermin, Hawking, and Carroll each promote an instrumentalist reading of science's proclamations about space-time. Note, however, that they all fall short of saying that what constitutes an acceptable scientific theory is a merely subjective matter. Note, in particular, Carroll's confident reference to the "real world." Reality is far from irrelevant in science, even from an instrumentalist standpoint. Dispensing altogether with the notion of an objective reality describable by science would make the predictive success of our theories describing the physical universe an inexplicable miracle. Reality, though only imperfectly grasped via observation and inference, constrains our selection of one theory over another; a scientific instrumentalist just concedes that our representation of reality is inescapably mediated by our perceptual faculties and ways of thinking. At the same time, some descriptions of the universe remain truly and objectively better than others because they explain more of what we see, or permit more accurate predictions about future observations.

In the non-quantum world the theory of relativity purports to describe, it is extremely well confirmed. A standard interpretation of relativity treats space-time as a real matrix for events. We may therefore say that space-time realism is a good assumption because it comports well with our observations. Can we go ahead and conclude that space-time is indeed real? We must retain modest expectations with regard to achieving a definitive answer to this question, no matter what advancements in our knowledge may be ahead.

One question we can fruitfully investigate further is the subsidiary (and for many, even more contentious) issue of the *passage* of time. Remember that, for the temporal realist Newton, the essential and definitive property of time is its "regular and equable flow." In post-Newtonian philosophy, much attention has been focused on this notion of the alleged movement of time. The passage of time is really the aspect of it that means the most to us in our day-to-day lives, and it is where the studies of the nature of time and the experience of time come together.

WORKS CITED IN THIS CHAPTER

Alexander, H. G., ed. *The Leibniz-Clarke Correspondence* (Manchester, UK: Manchester University Press, 1956).

Carroll, Sean. *From Eternity to Here* (Oxford, UK: Dutton, 2010).

Einstein, Albert. *Relativity: The Special and the General Theory* (London: Routledge, 2001).

Hawking, Stephen. *A Brief History of Time* (New York: Bantam, 1988).

Mermin, N. David. "What's Bad about This Habit," *Physics Today* 62, 2009, 8–9.

Minkowski, Hermann. *Address to the 80th Assembly of German Natural Scientists and Physicians*, September 21, 1908.

Newton, Isaac. *Principia Mathematica* (1687).

Other works of relevance to the issues in this chapter

Callender, Craig. "Is Time an Illusion?" *Scientific American*, June 2010, 59–65.

Chakravartty, Anjan. "Scientific Realism," *The Stanford Encyclopedia of Philosophy*, ed. by Edward N. Zalta, http://plato.stanford.edu/entries/scientific-realism/.

Dainton, Barry. *Time and Space* (Montreal: McGill-Queen's University Press, 2002).

Epstein, Lewis Carroll. *Relativity Visualized* (San Francisco: Insight Press, 1985).

Huggett, Nick, and Carl Hoefer. "Absolute and Relational Theories of Space and Motion," *The Stanford Encyclopedia of Philosophy*, ed. by Edward N. Zalta, http://plato.stanford.edu/entries/spacetime-theories/.

Chapter 4

Does Time Pass?

Classically, the definitive property of time is that it passes, or flows. The alleged passage of time is intimately bound up with the notion of a real past, present, and future, because the passage of time is the change that occurs as events go from being future to being present to being past. Is this essential characteristic of time (or such events in time) real?

REASONS TO THINK NOT, PART ONE (LOGIC)

The notion of a real, independent flow of time was mentioned by Newton, but it was not unprecedented. Plato, as we have seen, appeared to identify time with the motion of the heavenly spheres. Their rotations and the passage of time are one and the same. Fourth-century Neoplatonist philosopher Iamblichus of Syria also spoke of the flow of time from future to present to past—though he was aware of the Parmenidean paradox and consequently declared this temporal flow merely apparent.

The metaphor of a river of time is often employed to illustrate the idea of the alleged flow, or passage, of time. Even as a metaphor, this image is a little unclear. Are we on the riverbank, watching events roll by, or are we on the river, passing by events on the bank?

The distinction between these two visualizations probably doesn't represent any real difference. Correspondingly, the use of the terms 'flow' or 'passage' is ambiguous when applied to time: It can refer to how events seem to change, in that they could be described as progressing 'backward' from future to present to past; alternatively, we could say that changes in our experience depend on a sort of moving now—a unique present that constantly and fluidly moves 'forward' into the future.

Either way, the notion of the passage of time is closely tied to that of change over time. It was changes in existence, events, and properties over time that Parmenides had attacked as an illusion. Parmenides argued that the notion of changes over time presupposes a real future that becomes present and a present that becomes past (i.e., a flow or passage of time like the one alluded to by Newton). We have seen that some of Newton's ideas about time have been undermined by post-Newtonian developments. What have we learned in the modern era that helps us assess Parmenides' claims about the reality of change over and of time?

In 1908, about the time Einstein was publishing his original series of breakthrough papers on the relativity of motion, space, and time, English philosopher and temporal idealist J. Ellis McTaggart produced an extremely influential argument for the unreality of time that built on Parmenides' original argument for temporal idealism. Discussion of McTaggart's core argument against the reality of time has been an important focus for philosophers ever since.

Like Parmenides, McTaggart felt that the reality of time is predicated on the reality of change, where change is understood as involving the passage of time. Let us call the theory according to which change is a real aspect of events the **dynamic theory** of time. The dynamic theory really couldn't seem more natural or intuitive—even obvious! It incorporates the idea that there is something special about the *now*, the present moment. *This* moment is special, in that it picks out what is really going on, as opposed to what was or will be happening. It is also changing, in that, even as we speak, it is replaced by a new *now*. Future events are coming to pass, as what is happening now fades into the past. Time passes and—so to speak—there is no time like the present. This change in time is really what change is all about, and the reality of change is, it seems, undeniable.

McTaggart's argument against the dynamic theory is logical as opposed to empirical. It is both simple and devastating. As we have seen, there are two ways of understanding the ordering of events in time: in terms of 'past,' 'present,' and 'future,' and in terms of 'earlier' and 'later.' The former series (McTaggart calls it the **A-series**) is dynamic: Events are future, then they become present, and then past; these properties or determinations of events (futurity, presentness, and pastness), in other words, change over time. A-series change is what we measure with clocks. Events change with regard to whether they are future, present, or past, and they change with regard to the degree in which they are future or past (i.e., over time, they become less future or more past). The A-series theorist is thus committed to events having the intrinsic temporal properties of pastness, presentness, and futurity, which they lose or gain over time. As we can now see, a commitment to seeing time in terms of the A-series is integral to the position that the passage of time is a real phenomenon.

The other series (the **B-series**) refers to the sort of information you could get from looking at events recorded on a calendar: You would learn the dates of these events, and you would know which ones come earlier than others, which come later, which come on the same date. This kind of relation between events does not change over time; if an event is at any time simultaneous with, earlier than, or later than another, then it is so at all times. For the B-series theorist, genuine temporal properties have only to do with timeless B-series relations, rather than with any changeable properties, like presentness, that events have in themselves. In other words, the B-series theorist is a realist about time but an idealist about the passage of time (i.e., affirms the reality of B-series relations but denies the reality of A-series properties). This is the **static theory** of time.*

The A-series represents change in and over time; it is on the reality of A-series temporal properties that the reality of the passage of time rests. So the key question is "Are A-series temporal determinations real?" Are we justified in saying that some events are really present, some are really past, and some are really future? Wouldn't you be missing some really important information if you didn't know what time it is *now*?

Parmenides claimed that when we say an event actually is future, or present, or past, we are committed to saying that that event is future, present, *and* past. But these temporal determinations are incompatible. Ascribing each of these determinations to each event would also lead to other nonsense, such as the conclusion that Napoleon Bonaparte is both alive and not alive, and dinosaurs both walk and do not walk the Earth.

The natural response to Parmenides is that any actual event does have each of these contradictory determinations, but only at different *times*. At one time Napoleon is alive, and at another he is not; whether or not the statement "Napoleon is alive" is true depends on when the statement is made. After his death, we change the verb tense we are using, in describing his state of aliveness, from "is" to "was." Our use of tenses reflects our commitment to the objective reality of A-series properties, in that it is changes in the A-series properties of things that make changes in the tense of a declarative

---

* The dynamic theory is also sometimes called the **A-theory**, and the static theory is called the **B-theory**.

statement like "Napoleon is alive" appropriate. The truth of the resulting 'tensed' statement depends on the A-series situation (i.e., whether Napoleon is alive in the present, or only in the past). The dynamic theorist takes tensed statements to be basic and ineliminable. The **truth conditions** for a statement are the conditions that have to be in place for that statement to be true. A-series theorists are committed to tensed statements being made true by A-series facts. They are committed, in other words, for the truth of a statement like "X is the case" to depend on the fact that X is *present*.

Here McTaggart points out a problem with regard to the truth conditions for any statement that involves the use of tenses like 'was,' 'is,' or 'will be.'* Can the proponent of the dynamic theory explain *when* a tensed statement like "Napoleon is alive" is true? Take a future event "P" (say, the 2032 U.S. presidential election). Let us assume that the election will indeed take place, so the statement "P will happen" is true. But "P will happen" is not always true.** P will be present in 2032, so "P will happen" will not be true if uttered then. After 2032, of course, P will be past, so "P will happen" will not be true then either. When is "P will happen" true? "P will happen" is true when P is future *in the present*.

But this doesn't answer the question, says McTaggart. This reply doesn't tell you how the truth of "P will happen" should be assessed, because the statement "P is future in the present," although true now, is not true at other times. After 2032, P is past in the present: At that point, "P is future in the present" is false. So " 'P will happen' is true when P is future in the present" doesn't answer the question as to when "P will happen" is true. We need to know when "P is

---

* This is actually not quite the way McTaggart puts it, but has proven popular as a reconstruction of the essence of his argument.

** As you know very well if you are reading this book after 2032! For you, "P will happen" is false, whereas "P did happen" is true.

future in the present" is true; and the answer is that "P is future in the present" is true in the present.

Now the problem arises yet again at the next level. Specifying that "P is future in the present" is true in the present (but false in the future) won't be enough: Now you have to specify when the statement " 'P is future in the present' is true in the present" is true, and so on. This is what is known as an infinite regress. There is no way out; there is no way to fully specify the truth conditions for a tensed statement in A-series terms. You can only end the regress by bringing in B-series locators for the utterance and the subject of the utterance (e.g., "P will happen" is true *if uttered earlier than 2032*). Parmenides had claimed that there is a contradiction inherent in ascribing A-series properties to events, in that every event would have to be ascribed each contradictory property; McTaggart now shows that any attempt to explain that problem away by reference to changes, over time, in the truth values of our tensed statements will just continue to generate new contradictions ad infinitum.

We have defined the passage of time as the movement of events from future to present to past (or, alternatively, the movement of the present as the future becomes present and the present becomes past). Because talk about the past, present, and future gets us into an infinite regress, such talk is conceptually incoherent. McTaggart concludes that the notion of the passage of time is conceptually incoherent.

There are other odd consequences of supposing that certain events really have the property of being past, present, or future. Take some past event like the Battle of Waterloo. One would think that past events, being over and done with, should not be susceptible to further changes. Yet, as Ulrich Meyer points out, it is a consequence of the dynamic theory of time that something very odd is happening even now to the Battle of Waterloo: It is becoming *more past*. If we insist on the reality of A-series properties, this conclusion seems inevitable. But how can a past event be changing now? This is another reason to wonder, with Parmenides and McTaggart, if we are not systematically confused, in our day-to-day lives, about the reality of the passage of time.

One option explored by dynamist opponents of McTaggart is to say that the past and future don't exist. The position that time and change are real, but the past and future are not, is **presentism**. The presentist embraces one side of Parmenides' dilemma, wherein he said we talk about the past and future as both existing and not existing. The presentist embraces the notion that the past and future do not exist. Only the present moment exists. We live in an ever-changing 'now': The past exists, so to speak, only in our memories, and the future only in our imagination. As Parmenides had pointed out, it is natural, much of the time, to contrast the presence of the present with the nonexistence of the past and future: The past is 'over and done with,' and the future is 'not yet.' The present moment is special: We can be directly aware of what is happening now, unlike what happened in the past and what will happen in the future.* Only what is now is what is. So perhaps Parmenides and McTaggart are wrong to think that we are committed, if we believe in change, to treating events in the past and future as real. If so, then the contradictions McTaggart finds in the dynamic theory of time don't arise; what is true is simply what is true now, and not at other times—because other times don't exist.

---

* It is true that the information we are aware of at a given time is not entirely current, in that the limited speed of light and sound means that what we are now experiencing are events that happened some time ago. However, we can at least think of the arrival and processing of information as something that is happening, literally, *now*.

Yet we are committed, in our ordinary manner of conceptualizing time, to the reality of the past and future. According to the presentist, a past object, such as Napoleon's beloved horse Wagram, does not exist. Yet we seem perfectly able to speak meaningfully about Wagram the horse, such as when we describe its color and size. What are we referring to, if not to Wagram? The B-theorist can explain how we can refer to Wagram, because *Wagram exists in 1809.* The presentist has no easy answer to this: The (otherwise perfectly natural) response that Wagram *existed* in 1809, but no longer, exposes the presentist to McTaggart's concerns about an infinite regress of truth conditions for tensed sentences.

Consider further any true statement about the past, such as that the Berlin Wall fell in 1989. What makes this statement true? Presumably, a fact about what happened in 1989. The presentist can't consistently agree to that, though, because presentism doesn't allow for the reality of events that aren't happening now. The fact that people *now* have memories of the incident, or that there are *now* piles of concrete rubble where the wall is said to have been present, doesn't make it true now that the wall fell in 1989—or that it ever existed, for that matter. Rather, *only the event of its falling in 1989 makes it true now that it fell.* Insofar as we think that we can make true statements about the past, it must be false that only the present exists. This is one reason Parmenides said we are committed to the reality of non-present times (even as we are also, inconsistently, committed to their unreality).

Furthermore, presentism appears to be self-defeating as a defense of the possibility of change. Recall Zeno's arrow. The paradox of "The Arrow" stated that, at any instant, any object is at rest; if time is composed of such moments, then no object is ever in motion. The standard response is that time is not composed of infinitesimal moments, and motion is motion over stretches of time. But this answer is not available to the presentist. There are no other times, according to the presentist, so motion can only be a matter of what is happening now. Zeno's challenge showed that any adequate analysis of motion, or any other sort of change, must make reference to times other than the present instant. Thus presentism abandons change in an attempt to save it! That is why Parmenides thought that, if we believe in change, we are thereby committed to a real past and future.

The apparent failure of attempts to rescue the dynamic theory of time from the problem of conceptual incoherence doesn't mean that we have to conclude, with Zeno, Parmenides, Augustine, and Kant, that reality lacks any true temporal dimension. The alternative to the A-theorist's dynamic theory of time is the B-theorist's static theory, which rejects the objective reality of A-series properties, but embraces the reality of unchanging B-series temporal relations (i.e., earlier than, simultaneous with, and later than). According to the static theory of time, the fall of the Berlin Wall does precede, say, the 2032 U.S. presidential election. It always has preceded it, and it always will precede it. The real temporal properties of events do not change. Events do not come and go, yet they do exhibit temporal relations: relations, that is, of the B-series sort. So time order exists, according to the static theory, but dynamic change (i.e., change involving the passage of time) does not. The static theorist believes in change, but only understood in a way that does not commit one to the passage of time: *Change*, on the static theory, is to be understood as referring merely to the world being timelessly one way at one moment and timelessly another way at a subsequent moment.*

According to the static theory, past, present, and future are subjective, perspectival terms. This theory treats time a lot like space: There

---

* Tim Maudlin has suggested that "static" might not be the best word for this theory, because calling space-time static suggests yet another time dimension in which space-time is not moving. It is important to note that this is not what is intended by the term. Bearing this in mind, I think "static" effectively captures the desired contrast with a temporal passage view.

is nothing objectively special about temporal location, any more than there is about spatial location. The temporal location of "now" is just as much a subjective matter as the spatial location of "here." All events exist timelessly in an eternal, unchanging order (a view also called **eternalism**, in direct opposition to presentism). The advantage of this theory of time is that it avoids the contradictions inherent in the notion of temporal properties that change over time, while preserving our sense that events are ordered in time.

Eternalism brings with it an interesting view of the human person over time. From the eternalist viewpoint, a human being is not wholly present at any moment in time. Rather, we are extended in space-time, like four-dimensional worms, with only a part of us, or space-time slice, existing timelessly at any space-time point. As odd as this sounds, upon examination, this makes more sense than any account of personal identity over time derivable from the dynamic theory. The dynamic theory implies that we exist in our entirety at any time, even as time passes. This means that for any existing person, there is a present self, many past selves, and, potentially, many future selves. But they are all the same person, just at different times in his or her history. Each self at each time is a whole person, distinct from all those related selves at other times, and yet identical to them. This engenders another Parmenidean paradox, because it appears that the reality of the passage of time would mean we both are and are not identical to our selves at other times.

REASONS TO THINK NOT, PART TWO (PHYSICS)

The static theory is also sometimes called the "tenseless theory" of time. According to the static theory, our use of tenses (as in "Napoleon was alive," "Napoleon is dead," "This will happen," "That just happened") is useful to us as beings who operate from a particular perspective mediated by memory and anticipation, but does not capture anything inherent to the real world. Tensed sentences like these just describe what things look like from one's perspective at a particular tenseless moment: At one timelessly existing moment, I anticipate some event E; at some later, timelessly existing moment, I remember E. At some intermediate, timelessly existing moment, I experience E.

The non-presentist proponent of the dynamic theory—who doesn't want to deny the reality of the past and future as the presentist does—might reply by insisting that McTaggart and the static theorist are simply failing to 'take tense seriously.' From an introductory text on metaphysics by John Carroll and Ned Markosian:

> [T]he A-theorist's best move is to say that we must "take tense seriously." This means, roughly, that there is a fundamental and unanalyzable difference between saying that something is φ and saying that it *was* φ or *will be* φ. Taking tense seriously also means that propositions have truth values at times, and can change their truth values over time.

So tenses, on this view, pick out a fact about the world that is just not susceptible to the kind of analysis McTaggart would subject them to. There is no further question to ask as to when a statement like "Napoleon was alive" is true. Every time one asks about how to understand a tensed statement, this A-theorist just refuses to answer; there is nothing more to be said about the significance of a "was" or "will be." So the vicious infinite regress of specifying truth values can't get started.

This is not the end of the discussion, however, because there is a very significant contribution to this debate to be made from empirical science. Recall that Einstein's well-confirmed theory of the relativity of motion makes simultaneity a relative matter: Two events that are simultaneous to an observer in one inertial reference frame will not be to a different observer moving relative to the first one. We saw this in the case of Isaac and Albert, who each have their own planes of simultaneity, encompassing those events that are simultaneous in their frames of reference. And a third observer, in a third reference frame, will have a different assessment than either of them. Different observers may represent the universe of events differently in space-time. Because there is no absolute motion, no observer has a privileged vantage point to determine which events are really simultaneous and which are not. This means that the set of events 'present' to Isaac is different from Albert's set, and each is equally justified in identifying his set as present. This also means that neither would be justified in identifying a particular moment that everyone should agree is 'now,' the present moment. The same goes for pastness and futurity: An event Isaac considers past may be present, or even future, for Albert, depending on its relative distance and their relative velocity (see figure 4.1).* If there is no privileged vantage point from which to determine the 'truth' of the matter—and the whole point of relativity is that there is not—then temporal properties like past, present, and future cannot possibly be aspects of reality as it is in itself. They *must* be subjective and perspectival in nature.

As strange as it sounds, if relativity is right, then the dynamic theory of time must be wrong. All events, past, present, and future, are present from some frame of reference. Without a real past, present, and future, there can be no passage of time and no dynamic change.

---

* Figure adapted, courtesy of Wikipedia user Acdx, based on file: Relativity_of_Simultaneity. svg. Accessed Oct. 22, 2012, at http://en.wikipedia.org/wiki/File:Relativity_of_Simultaneity_ Animation.gif.

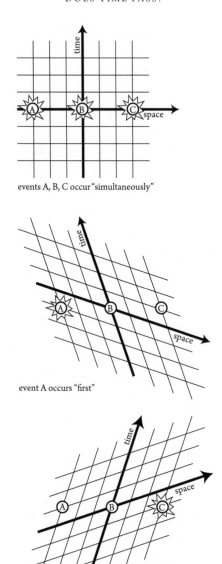

events A, B, C occur "simultaneously"

event A occurs "first"

event C occurs "first"

Figure 4.1. The perceived orientation of space-time, and thus the perceived order of events, depends on one's relative velocity; this means that what events are *present* is also a relative matter.

Does this mean Parmenides was right all along? Actually, the logical and physical contradictions inherent in the notion of dynamic change only tell half the story. Parmenides denied not only the objective reality of change, but also the very existence of time itself. As we saw in the last section, there is a theory—the static theory of time—that rejects the objective passage of time, but preserves objective temporal relations between events. Does the relativity of simultaneity imply that the static theory of time is also wrong?

The static theory proponent maintains that B-series relations are fixed and unchanging—temporal properties are what they are 'eternally,' so to speak. Relativity permits observers in every reference frame to agree on the temporal order of events within any particular light cone. Earlier-later relations remain fixed within any reference frame. This permits us to speak of fixed time orders of events, even when taking relativity into account.

How should we represent or imagine a world with time-ordered events, yet without the passage of time? Static theorists typically refer to the world of unchanging events in space-time as a **block universe**. This refers to the notion of a four-dimensional block depicting everything happening in space over time. Observers in different reference frames will produce different representations of the block; that is to say, they will locate events in space and time somewhat differently. In a Newtonian block universe, all events would exhibit a set of B-series relations that are the same for all observers. In contrast, that is only true in Minkowski space-time for any given frame of reference. The complicating factor is that spatiotemporal distances are relative, so different observers will locate the light cones emanating from different spatiotemporal locations differently. For example, take observers Isaac and Albert, moving rapidly relative to one another. Isaac orders a series of events A-B-C-D (within one light cone) and a series of events E-F-G-H (spatially separated from A-B-C-D) as follows:

A & E occur simultaneously,
then B & F occur simultaneously,
then C & G occur simultaneously,
then D & H occur simultaneously.

But Albert may, with equal authority, locate them as follows:

A occurs,
then B occurs,
then C & E occur simultaneously,
then D & F occur simultaneously,
then G occurs,
then H occurs.

This illustrates how observers in different reference frames must disagree about ordering with regard to spatially separated events, yet can agree on a temporal ordering for merely temporally separated events. This implies a fixedness with regard to time relations within a given light cone; and, because causality can only function within light cones (since nothing travels faster than light), *from the standpoint of a physical description of the operation of natural laws, the earlier-later relation is meaningful and useful.*

The designation of something as past, present, or future is only meaningful from the subjective perspective of a particular observer. However, as we have seen, relativity permits us to state objective B-series relations between any causally relatable events. It is in this sense that, despite the relativity of simultaneity, the static theory of time can speak of an invariant spectrum of events, and thus in terms of fixed, true earlier-later relations.

Certainly, the terms 'past,' 'present,' and 'future' are useful in that the attribution of beliefs using these terms can be used to explain and predict human behavior. But it appears they have no role to play in an objective scientific representation of the distribution of events in space-time. Take someone drinking a cup of coffee. Explaining her action involves, in part, ascribing to her the belief that her cup of coffee is *here*. She reaches across the table for some sugar. Explaining that action involves, in part, her belief that the sugar is *there*. But 'here' and 'there' are subjective designations; for someone else, the sugar might be *here* and the coffee over *there*. There is no objective, third-person standpoint, in respect to which the coffee cup is truly here or there. Just like 'here,' *'now'* is a term that has no application in a description of the world that excludes people's subjective beliefs and attitudes. To include 'now' in that description would presume an absolute present that Einstein's theory dispenses with when it dispenses with absolute simultaneity. (Furthermore, if and when the theory of relativity is superseded by a more complete theory, this core element will likely remain.) McTaggart agrees, finding a logical incoherence in the notion of pastness, presentness, and futurity. The result is a universe of all events "past, present, and future" having equal status at all times as past, present, and future—which is to say, having no such status at all. All events just BE, related only by BEing earlier than, simultaneous with, or later than each other.*

---

* I don't mean to suggest that there is a complete consensus on the static theory of time among philosophers of time. A few distinguished philosophers think the passage of time is compatible with a block conception of the universe. See, for example, Tim Maudlin's *The Metaphysics within Physics* and Michael Tooley's *Time, Tense, and Causation*. In Maudlin's case, however, I am not sure he is really talking about temporal passage so much as temporal asymmetry, or directionality, which would be a different matter (see chapter 5).

THE PERVASIVE SENSE OF PASSAGE

Some philosophers claim that there must be something to dynamic change and the passage of time because we have a distinct *sense* of the passage of time. Peter van Inwagen refers to the "sensation of temporal motion," and Donald Williams to the "felt flow of one moment into the next." If there really were such a sensation, it would constitute direct empirical evidence for the passage of time.

I am dubious that this sensation truly exists. However, I think I understand why people might be led to make this sort of report. We do not typically perceive processes as a succession of atomic units. Our phenomenal experience of motions and other continuous processes has a certain sensible quality that does not seem to be captured by the notion of our having successive beliefs about part of the motion or process being future, present, or past. As we discussed in chapter 2, seeing a clock's second hand move is not the same thing as seeing it occupy, or remembering it occupying, different locations at different times. Listening to a melody is different from just hearing the different notes as occurring in sequence, or remembering the notes as having occurred in sequence. This was Kelly's phenomenon of "pace perceived." I would suggest that it is a misinterpretation of this phenomenon that gives some people the notion that they literally feel the passage of time.

Kelly views the explanation of this phenomenon of pace perceived primarily as a challenge for neuroscience, and I agree. Somehow the brain converts sensory input into "experiences as of continuous, dynamic, temporally structured, unified events." It is thanks to this that our experience flows in the sense that the perception of movement or change is not experienced as an accumulation of static moments, but as an ongoing continuity. The experience of continuity is not itself a distinct *feeling* of flow, but it can explain why one might be inclined to report such a feeling.

My view, then, is that the alleged felt passage of time does not require any corresponding objective phenomenon along the lines of Newton's "flow" of absolute time. It would be difficult in any event to understand what a real flow to time would amount to, or how such a thing could be sensed by us even if it did exist.

This conclusion is bolstered by further reflection on the very notion of a real passage of time. The image of a flow of time implies a sort of movement: Events move into the past, and we observers move into the future. In every other case, when we talk about movement or change, we are talking about movement or change *over time*. But what about the idea of time itself as passing? Time is passing relative to what? How could time pass with respect to itself? Further, if time passes, at what *rate* does it pass? Normally, we think of things in a state of change or motion as having a rate of change or motion measured according to, say, per hour, per minute, per second. How would the passage of time be gauged? "One second per second" is not a rate: It communicates no information.

And yet, for all the arguments of philosophers and physicists, the notion of giving up on dynamic change and the passage of time seems not only wrongheaded but impossible. The primary reason for this is the way the difference between past, present, and future is wrapped up in deep-seated psychological and emotional attitudes that are universal to human beings. For example, the static theory entails that a deceased loved one is 'always' alive at an earlier time. But knowing that the static theory of time is true doesn't make the loss of a loved one any less painful. Albert Einstein pretended otherwise. Speaking at the funeral of his old friend Michele Besso, Einstein said:

> Now Besso has departed from this strange world a little ahead of
> me. That means nothing. People like us, who believe in physics,

know that the distinction between past, present, and future is only a stubbornly persistent illusion.

And yet, as much as Einstein may have intellectually grasped the truth of this view about time, could he truly have derived much comfort from it upon the loss of a friend? If so, he was surely one of the only people ever to do so. The disconnect between what one knows and what one feels, in this instance, is striking. There is an essential asymmetry to how we treat the past and the future. We fondly remember (or regret) past events, while we eagerly anticipate (or fear) future events. There doesn't seem to be any way, furthermore, to translate expressions of such attitudes into 'tenseless' B-series talk. Arthur Prior, the important New Zealander logician, made this point as follows:

> One says, e.g., "Thank goodness that's over!", and not only is this, when said, quite clear without any date appended, but it says something which it is impossible that any use of a tenseless copula with a date should convey. It certainly doesn't mean the same as, e.g., "Thank goodness the date of the conclusion of that thing is Friday, June 15, 1954," even if it be said then. (Nor, for that matter, does it mean "Thank goodness the conclusion of that thing is contemporaneous with this utterance." Why should anybody thank goodness for that?)

It seems nonsensical to suggest that our statements involving past, present, and future could or should really be rephrased in terms that don't involve changing temporal determinations. This would require us, among other things, to admit that our feelings of relief, regret, fear, and hope are entirely misplaced. In Kurt Vonnegut's *Slaughterhouse Five*, the protagonist comes to experience his whole life timelessly; he begins to appreciate that all the events in his life simply BE, rather than come to be and pass away. This actually instills in him a sense of peace in thinking both about missed opportunities during the early parts of his life and his own death at the other (timeless) end of his existence. But life isn't like this. This is not the way we actually experience things, so having different attitudes with regard to past and future events seems inevitable.

So denying the reality of change involves denying the legitimacy of what would otherwise seem to be perfectly normal emotional attitudes; some have gone further by arguing that denying change is literally incoherent. Israeli philosopher Yuval Dolev argues that our attitudes and sentiments are essential to fixing the *meaning* of 'past,' 'present,' and 'future,' so asking whether attitudes toward past, present, or future conditions are rational or appropriate is incoherent: According to him, the future, for example, just is that which we have an anticipatory attitude about. If so, then anticipation can't be wrongheaded in the way the static theorist might claim. Dutch philosopher P. J. Zwart claims that even questioning the reality of the passage of time is self-defeating: How, he asks, can the question itself be articulated except in time? Can any question be meaningfully asked or understood without presuming the passage of time from the inception of the question to its conclusion?

Even if you wanted to deny dynamic change, Michael Dummett points out that you still would need to explain the *illusion* of change, in the form of changing attitudes toward events according to whether they are perceived as future, present, or past. But how can you explain the illusion of change except by reference to changing apprehensions on the part of the victim of the illusion? Then you have change again: If you can't explain the illusion of change without reference to changing apprehensions, then you haven't successfully taken change out of the equation.

As we have noted, change itself takes on an attenuated meaning in the context of the static theory. In the block universe of the static theorist, change just means that the world is in one state at one moment and in a different state at a later moment. This certainly is not how we ordinarily think of change. We ordinarily think of change as involving the absolute becoming of new events and states of affairs, and the passing away of old ones. If time does not pass, how can anything *happen*? McTaggart thought that what the B-theorist calls change is more like spatial variation (i.e., change of spatial location) than temporal variation; this is part of the reason he was forced to conclude, with Parmenides, that change is unreal.

Accepting the static theory would also mean that we would have to reevaluate how we think about causation, natural laws, and scientific explanation. We normally think in terms of causes "bringing about" their effects, of effects "following" their causes. Consider a law of nature, such as Newton's third law of motion: *For every action there is an equal and opposite reaction*. Without dynamic change and the passage of time, we are left only with an eternal, static earlier-later relationship between so-called causes and their effects. The term "reaction," in Newton's law, seems to imply a dynamic change as the action *produces a reaction that follows upon it*. But on the static theory, the effect simply, timelessly, exists later than its cause. Then how can we still think in terms of causation, of 'bringing about'? The action (A) and the so-called "reaction" (B) just sit there, timelessly, next to each other; there may be a strong statistical correlation, in that similar actions and similar reactions are reliably found sitting side-by-side elsewhere on the changeless timeline—but, other than that association of adjacency, what is the *connection* between such events? To say that B's presence requires A's earlier presence is just to raise anew the question of how a B can require an A, if A isn't *bringing B about*. On the static theory, the notion of a real connection, a causal link, between A and B seems not only mysterious but incoherent.

Scientific exploration consists, to large degree, in an attempt to uncover the natural laws that govern the phenomena that interest us. Scientific explanation, in turn, refers to these laws in accounting for these phenomena. A simple natural law may take a form like *Events of type P are always followed by events of type Q*. We then explain why an event of type Q has taken place by noting the prior incidence of an event of type P, plus the rule that one must always follow the other. The problem is that a statement of this form only qualifies as a law if we also believe something along the lines of *Events of type P cause events of type Q to happen*—that is the difference between a law and a coincidence! Yet on the static theory, all we can say is that events of type P BE (timelessly) statistically well correlated with later events of type Q. The notion of events being brought about by other events seems to have no place in a static worldview, so neither does the notion of a natural causal law.

Scientific explanation involves notions of causes and causal laws at its core. If we take the static view of things seriously (and it seems that we ought to), then events that are causally connected in our explanations are really just timelessly, statistically associated with each other. Without the notion of a present event that *brings about* a merely potential, future event, are we left with a notion of causation robust enough to ground any claim to understand fundamental laws as such? If causation isn't real, then neither, it seems, are laws of nature.* Our attempts to explain the phenomena we encounter

---

* For both Hume and Kant, this result would be viewed as a feature of the static theory, rather than as a bug. Hume famously noted that the alleged causal connections between events are not actually part of our experience of those events. He attributed our belief in causal connections to the feeling of expectation we get when we have become accustomed to types of events occurring together. Judgments about the reality of such connections outstrip what is found in experience. Kant the idealist thought that causation is fundamentally a matter of how we organize the raw data of experience, so the concept of causation applies only to the world considered from the human perspective, as opposed to the world as it is in itself.

reduce to mere descriptions of inexplicable correlations of events. In that way, the static theory implies that reality is unknowable.

The static theory's strange implications for causation and natural laws are mirrored by its implications for the closely related concept of probability. Ordinarily, we would think that the future is governed not only by causation, but also by probability. There is a certain probability that it will rain tomorrow, or that the Dallas Cowboys will win the Super Bowl next year. But if the static theory is correct, then these events (the event of it raining or not raining tomorrow, or the event of the Cowboys winning or not winning the Super Bowl), timelessly BE. Either it BE raining tomorrow or it BE not raining tomorrow. Keep in mind that, as Einstein showed, what one person calls "tomorrow" is today in another reference frame. There is no question of any event being somehow merely potential or undetermined. Everything, at all times, just BE the way that it BE. So any judgment as to the probability of a so-called future event is just a statement as to our limited perspective on timeless reality.

No wonder the static theory seems counterintuitive. The way we think about the world inextricably involves change, the passage of time, anticipation, regret, causation, and probability. Yet this persistent commitment to change has been called an error and confusion since the time of Parmenides. This is the core challenge in the contemporary philosophy of time: how to reconcile the seeming ineliminability of the experience of the passage of time (**manifest time**) with the cold, hard conclusions of logic and physics (**scientific time**). As we shall see in the next section, although there is work to do to achieve this reconciliation, we may be able to sketch the outlines of an eventual solution. It turns out that Kant and Darwin are equally helpful in understanding our commitment to temporal passage.

TEMPORAL PASSAGE AS ADAPTIVE PSYCHOLOGICAL PROJECTION

The solution I would like to propose* is that A-series change and the passage of time are mind-dependent in the sense of being merely matters of **psychological projection**; however, it is a projection of a special kind, in that it is conceptually indispensable to any coherent representation of the world. To say a manner of representing the world is indispensable is to say that no genuinely thinkable description of the world without it is possible. This, I shall suggest, gives us a certain entitlement to include passage in our description of the world, without contradicting the conclusions of science and logic that the passage of time is a merely subjective phenomenon.

'Psychological projection' refers to a loose collection of phenomena in which one represents an internal, subjective feeling or sensation as an objective feature of the world—or, more broadly, in which one's commitment to some state of affairs being the case is explained by psychological facts about oneself, rather than by the world really being that way. An example of projection is our everyday attribution of color properties to things around us, say, a green umbrella or the blue sky. 'Greenness' is not really a property that the umbrella has; rather, the umbrella is reflecting light at a certain wavelength, and our brains are constituted to respond to that wavelength of light in a certain distinctive way. Greenness is a feeling, not a property of things. It is similarly so when we use adjectives like "loud" or "soft": Things can seem loud to us or engender a soft feeling, but, strictly speaking, it is nonsensical to propose that something literally *is* loud or soft. In contemporary philosophical

---

* Some of the material in this section is derived from my essay "Time-Awareness and Projection in Mellor and Kant," *Kant-Studien* 101 (2010).

parlance, properties like greenness, loudness, and softness are called "secondary properties." Such properties are not real properties of things around us, nor are they mere illusions or falsehoods: They are predictable responses to something real. Our attribution of such properties to the things causing these responses is a case of projection, in which a mind-dependent sensory response is understandably, but erroneously, attributed to things independent of the mind. As we have seen, arguments are stacking up to the effect that dynamic change isn't really a property of events around us. To explain why we so characteristically represent things around us as undergoing dynamic change, would a comparison to the projection of secondary sensory properties be helpful?

Philosopher Richard Gale says no, because A-series temporal determinations are not sensible qualities like color or sound. True enough, but Robin Le Poidevin replies that the perception of passage might be fruitfully compared to other sorts of projection, such as the projection of non-sensible qualities like virtue and vice. There is no real characteristic, describable by natural science, corresponding to what we call virtue; there are only sympathetic emotions that arise when we see patterns of behavior that we associate with pleasant results for ourselves or others. The ascription of virtue as a characteristic to persons is thus an example of psychological projection in the broader sense mentioned above. D. H. Mellor is another who endorses the notion that the experience of passage is a kind of projection. On his account, we experience different beliefs at different times about the presence (or pastness or futurity) of some event. It is the simple difference in these beliefs over time that leads us to represent events as, objectively, changing their A-series determinations. In other words, differences in one's own beliefs are projected onto the world in the form of ascriptions of objective change.

Mellor's account is consistent with the simple conclusion that a belief in objective change is a flat-out illusion and mistake. This doesn't sit well with our intuitions, though; it would be rare to find someone (even a philosopher or a physicist!) who lacks a deep-seated conviction that there just has to be something to the idea of change and the passage of time, no matter what philosophy and physics might say. We can look to Kant to find a hint at an answer to this conundrum. In chapter 2, we saw how Kant argues that the concept of temporal succession is presupposed in any coherent experience. His central insight is that we have to impose a temporal structure on our own experience in order to make any sense of it. He starts with the claim that we do not perceive time order directly; as his fellow temporal idealist Augustine also emphasized, at any given moment, all we have is our present perception and perhaps some present recollections. Recollection, as we have discussed, means nothing without a grasp of the difference between past and present. Coherent experience requires an innate ability to organize the world temporally, as opposed to originally deriving the very concept of time from our experience (i.e., finding it in the world).

When Kant talked about organizing our experience in terms of temporal succession, he did not mean change of A-series properties; the static theory of time allows succession, albeit in static B-series terms. This distinction was not a primary concern, for him. However, his theory suggests, in two respects, that the necessary conditions of human cognition involve conceptualization specifically in terms of dynamic change and the passage of time.

First, complex representation typically involves a process of building up that representation over time, where a sense of the incremental construction of the representation is essential to the process. From Kant's *Critique of Pure Reason*:

Without consciousness that that which we think is the very same as what we thought a moment before, all reproduction in the series of representations would be in vain. For it would be a new representation in our current state, which would not belong at all to the act through which it had been gradually generated, and its manifold would never constitute a whole[...]. If, in counting, I forget that the units that now hover before my senses were successively added to each other by me, then I would not cognize the generation of the multitude through this successive addition of one to the other, and consequently I would not cognize the number. (Guyer/Wood, trans.)

The activity of, say, counting means something only by virtue of the representation of one's activity as part of an ongoing process, where what part of the process is past matters to the meaning of the current phase. And counting, for Kant, is here just standing in for any generation of coherent experience out of an intrinsically inchoate succession of thoughts and experiences.

Second, as we also discussed in chapter 2, Kant makes the case that experiences do not come temporally presorted, and that the basic rules for doing the sorting are innate; further, he argues that we make temporal sense of our own experiences by relating them to things and events outside ourselves. Doing this requires that we have an instinctive grasp of the difference between future, present, and past, for the reason that—and this is the key point—we could never get started with sorting out our experience unless we presume that we are restricted both spatially and temporally with regard to what we can be perceiving at any moment. In recollection, I remember both the cause and the effect. Did the experience of the effect come before the cause, or after? We represent ourselves instinctively as experiencing restrictions on perception based on the spatiotemporal availability of the things we perceive: We automatically organize our experiences as being restricted to the here and now, with future and/or distant events inaccessible until we are in a position, in terms both of external conditions and of our own spatial and temporal location, to have the experience in question. *Without the presumption of such restrictions, there would be nothing constraining how we might organize a given set of perceptions.* A grasp, on some level, of the spatiotemporal conditions that make the difference between perceived and unperceived is thus necessary for coherent experience. As Kenyan philosopher Quassim Cassam puts it:

> To get a grip on the idea of perception of what can also exist unperceived, one must think of perception as having certain spatio-temporal enabling conditions, such that in order to perceive something one must be appropriately located—both spatially and temporally—with respect to it [...] [this assumption] enables one to account for the fact that a perceivable object is not actually perceived by appealing to the possibility that the spatio-temporal enabling conditions of perception are not met with respect to it.

This is not the same thing as representing oneself as being timelessly located in one place at one moment and in another place at another moment; in other words, the static theory of time is insufficient as a precognitive organizational principle. Rather, this picture involves essential reference to a dynamically changing location in time (and space) that explains the gradual unveiling of hitherto inaccessible objects or events, and thus includes the representation of change specifically in dynamic theory terms.

Kant's claim that experience is only possible by virtue of its organization in terms of rules involving a world of objects and events thus implies that one must innately presuppose that one's experience derives from dynamic changes in one's environment and spatiotemporal position; in short, one *must* begin by thinking in terms of having present experiences with further, potential experiences on the way. Otherwise, one could never get started with putting one's experiences together coherently, or sorting them out in a usable manner.*

Note that this doesn't mean that dynamic change really happens: It only means that change is an indispensable part of our basic conceptual scheme, a necessary conceptual presupposition that makes coherent experience possible. As with the case of virtue and vice, our attribution of change to events in the world is a matter of projection—though in this case, this way of representing the world is necessary to our having any coherent representation of the world at all. What this all means is that, whatever the world is 'really' like, we can't do without thinking of it in terms of dynamic change...which doesn't really come as a surprise to anyone who has tried (and failed) to imagine the world without it!

The conceptual indispensability of this manner of representing the world can be easily explained by its adaptive advantages. It would be surprising to find any species capable of conceptualizing the world in this way that failed to do so. Clearly, for any sort of being who acts on the basis of beliefs, taking the right actions at the right times is dependent on having beliefs like "*now* is the time to eat/sleep/resist/run away."

---

* This theory proposes that dynamic change is presupposed in the interpretation of experience, as a condition of any coherent experience. This conclusion must remain tentative, because (as we discussed in chapter 2) empirical investigation may show, say, an origin of the concept of succession in experience via a direct experience of change in a unitary, temporally extended act of awareness (i.e., a 'phenomenological present' of perception). I view Kant's a priori reasoning, in other words, only as providing a plausible hypothesis for empirical investigation.

James Maclaurin and Heather Dyke, philosophers in New Zealand who have been thinking about time-awareness through the lens of evolutionary biology, have argued that 'tensed emotions' like apprehension and relief are evolutionary adaptations, or side effects thereof. As they explain, a feeling of relief like the one Prior cited exists as a human emotional state because "at the time at which a fearful experience is past, from our temporal perspective, we *then* no longer have to expend great amounts of adrenaline trying to avoid it. It is this contrast which we interpret as relief." Conscious beings derive tremendous advantage from the ability to plan ahead and engage in temporally extended projects. The conscious pursuit of goals requires reflective planning; future-directed emotions like desire, anticipation, and apprehension are essential to the agency involved in planning and executing projects aimed at some later goal. All events being equally real (as the static theory insists) doesn't mean that all events deserve the same emotional response every time we think about them; it is essential to our prospects of survival that we respond differentially to events according to their causal relation to us at any given time, and we should expect this to be reflected in our emotions.

We can now understand the phenomenon of attitudes like Prior's "thank goodness that's over" in the context of the static theory of time. All that is really going on is that at 2:15 p.m., I am apprehensive about my imminent tooth extraction; at 2:30 p.m., I am distressed about the pain as my tooth is extracted; and at 2:45 p.m., I am relieved that the experience is over. That's it. This account includes a sequence of appropriate, time-dependent emotional states, without having to be committed to one of these times actually being *now* and the others as *past* or *future*. The absence of any absolute *now* is perfectly consistent with my having beliefs, at any moment, as to what is 'currently' happening, and as to what lies in the past or future. In fact, I need such beliefs to function. Needing certain beliefs in order to function is not the same as having the corresponding representation of the world constitute an accurate reflection of reality. What time it is now matters, though only *to me*, and only to me *now*; which is to say that, from the universe's perspective, it doesn't matter at all. Prior is right to point out that we cannot simply translate "Thank goodness that's over" (uttered on Friday, June 15, 1954) into "Thank goodness the date of the conclusion of that thing is Friday, June 15, 1954." These mean different things, but the first locution says something about change only insofar as we are bound to conceptualize it.

The question is whether this understanding of the conceptual indispensability of temporal passage helps resolve the problem of whether or not to accept the logical and scientific arguments against the dynamic theory of time. The idea that change is both conceptually indispensable and, apparently, unreal places us in an odd position. On the one hand, logic and science seem to leave no room for dynamic change; on the other, the conceptual indispensability of change would mean that we literally cannot contemplate a world without it. One might well argue that you can't have it both ways and say that change isn't real, but a changeless world cannot be contemplated: If you say change isn't real, are you not thereby contemplating a changeless world? This brings us back to the question we raised at the end of the last chapter, about the goals of science vis-à-vis capturing reality. There are a couple of ways we could try to resolve the conundrum at hand. We could conclude that we are simply incapable of grasping reality as it really is, and so can never penetrate to the true nature of things beyond formulating some mathematical models that yield successful predictions but are otherwise meaningless to us as representations of reality. Or, we could elevate the world 'as we know it' (i.e., including change) to the level of reality and reduce science's role to merely describing our manner of representing the world. Each approach has its drawbacks. On the first approach, we admit that there will always be a gap between our understanding of the world and the true natures of things. But the second approach is worse: On it, we downplay the distinction between the way things are and the way they appear to be, which would seem to make a mockery of the purpose of inquiry (i.e., finding the truth).

Because the alternative is to kill the patient in order to cure the disease, our conclusion to the issue of reconciling our perception of time with the logic and science of time must be to accept that there will never be some sort of final unification of experience and reality. Logic and science will tell us one thing about reality, even as our perceptual and cognitive faculties shall—and must—insist otherwise.

WORKS CITED IN THIS CHAPTER

Bardon, Adrian. "Time-Awareness and Projection in Mellor and Kant," *Kant-Studien* 101 (2010), 59–74.

Carroll, John W., and Ned Markosian. *An Introduction to Metaphysics* (Cambridge, UK: Cambridge University Press, 2010).

Cassam, Quassim. *Self and World* (Oxford, UK: Oxford University Press, 1999).

Dolev, Yuval. *Time and Realism* (Cambridge, MA: The MIT Press, 2007).

Dummett, Michael. "A Defense of McTaggart's Proof of the Unreality of Time," *Philosophical Review* 69 (1960), 497–504.

Gale, Richard. *The Language of Time* (New York: Routledge, 1968).

Kant, Immanuel. *Critique of Pure Reason* (1787).

Kelly, Sean. "The Puzzle of Temporal Experience," in *Cognition and the Brain*, ed. by Andy Brook and Kathleen Akins (Cambridge, UK: Cambridge University Press, 2005).

Le Poidevin, Robin. *The Images of Time* (Oxford, UK: Oxford University Press, 2007).

Maclaurin, James, and Heather Dyke. "Thank Goodness That's Over: The Evolutionary Story," *Ratio* 15 (2002), 276–292.

Maudlin, Tim. *The Metaphysics within Physics* (Oxford, UK: Oxford University Press, 2007).

McTaggart, J. M. E. "The Unreality of Time," *Mind* 17 (1908), 457–474.

Mellor, D. H. *Real Time II* (Routledge, New York, 1998).

Prior, A. N. *Papers on Time and Tense* (Oxford, UK: Oxford University Press, 2003).

Tooley, Michael. *Time, Tense, and Causation* (Oxford, UK: Oxford University Press, 2000).

Van Inwagen, Peter. *Metaphysics* (Boulder, CO: Westview Press, 2002).

Vonnegut, Kurt. *Slaughterhouse Five* (New York: Dell, 1991).

Williams, Donald. *Principles of Empirical Realism* (Springfield, IL: C. C. Thomas, 1965).

Zwart, P. J. *About Time* (Amsterdam: North-Holland, 1975).

Other works of relevance to the issues in this chapter

Callender, Craig. *Introducing Time* (New York: Totem Books, 1997).

Dainton, Barry. *Time and Space* (Montreal: McGill-Queen's University Press, 2002).

Huggett, Nick. *Everywhere and Everywhen: Adventures in Physics and Philosophy* (Oxford, UK: Oxford University Press, 2010).

Hume, David. *An Enquiry Concerning Human Understanding* (1748).

Kail, Peter. *Projection and Realism in Hume's Philosophy* (Oxford, UK: Oxford University Press, 2007).

Meyer, Ulrich. *The Nature of Time* (Oxford, UK: Oxford University Press, forthcoming).

Prosser, Simon. "Could We Experience the Passage of Time?" *Ratio* 20 (2007), 75–90.

Turetzky, Philip. *Time* (New York: Routledge, 1998).

Chapter 5

The Arrow of Time

If the dynamic theory of time is wrong, then the future isn't coming; the present isn't passing; and the past hasn't passed. But then why does it seem that time has a direction—that there is a difference between earlier (backward in time) and later (forward in time)? Could the difference between earlier and later also be just a subjective matter?

TIME'S DIRECTION

The perceptive reader will note that the rejection of A-series temporal properties (past, present, and future) raises a deep question about B-series temporal relations (earlier and later). We have discussed how our intuitive views about change can be understood within the static theory of time. The static theory embraces the block representation of the universe, according to which all times (or events) are equally real. But the static theory, by embracing the relational temporal properties 'earlier' and 'later,' adds a *directional ordering* to the block of events. The rejection of the dynamic theory means that time does not 'flow,' in the sense of movement or passage; nevertheless, even on the static theory, events are to be ordered from earlier to later. The very nature or meaning of the relations 'earlier' and 'later' includes the notion of a temporal direction or asymmetry. This asymmetry is easy to explain if we believe in a real past, present, and future: Event A is earlier than event B if it lies in B's past. But we have seen that there are good reasons to reject the notion of intrinsic temporal properties like past, present, and future. So what explains this one-way ordering of events from earlier to later, if it does not rest on the difference between past and future? This problem led McTaggart to question not only the reality of the passage of time, but also the reality of time itself. Is the one-way direction implied by the earlier-later relationship also an illusion or projection, or does the apparent directionality of time really correspond to something in nature?

The directionality of time is often referred to as the "arrow" of time (not to be confused with Zeno's paradox of "The Arrow"). In fact, there are a number of temporally asymmetric processes, each of which could be called the basis for the seemingly inherent directionality of time. The three most familiar are the psychological, the thermodynamic, and the causal.

THE PSYCHOLOGICAL ARROW

The **psychological arrow** just refers to the familiar fact that we remember (and never anticipate) the past, and anticipate (but never remember) the future. This fact explains how we come to have those asymmetrical temporal emotions discussed earlier, like sorrowful regret and eager expectation; we 'look forward to' what we anticipate, and we 'look back on' what we remember. We know things about the past but not about the future; we feel we can affect our future but cannot change our past. There is a perfect correlation between the psychological orientation of our lives and the direction of time. Yet surely the direction of time is not itself explained by this psychological asymmetry. That would make minds the cause of the order of time, which, in short, would make the difference between earlier and later an entirely human phenomenon; not only that, but it would seem also to make the difference between earlier and later dependent on what falls into the contents of each individual's memory. Clearly, it's the order of events that explains the psychological order, not the other way around. If we want to maintain that the directionality of time is a real phenomenon, then we need to look elsewhere to explain it.

THE THERMODYNAMIC ARROW

The **thermodynamic arrow** refers to the tendency of systems to move from overall orderliness to overall disorderliness; this is also referred to, in statistical mechanics, as the movement of a system to a higher state of **entropy**. Heat, like any other sort of energy, tends to dissipate, as does a drop of ink in water. Omelets do not become whole eggs; smoke, ash, and heat from a fire never reassemble themselves into wood. Why is this? It is because there are very many more ways for a system to be disordered than for it to be ordered. There are many more ways for the drop of ink to be diffused throughout the water than for it to be concentrated in one area. (Every distribution of the ink in the water is equally unlikely; it is just that most of those unlikely distributions involve the ink being spread out.) Simple chance thus dictates that, over time, any orderliness that we do not spend energy to create or maintain will tend to transition into a much, much more likely state of disorder.*

---

* Another way of putting it is to say that systems tend to move from a non-equilibrium state to an equilibrium state—for example, water poured in a pan tends to spread itself around nicely on the bottom of the pan rather than bunch up in some areas. This roughly equal distribution constitutes a more 'disorderly' arrangement in the thermodynamical sense.

At first glance, the thermodynamic arrow might appear to be the right kind of thing to account for time's direction: The seeming directionality of time, one might claim, derives from the fact that overall energy at 'later' times is always more diffused than at 'earlier' times.* We make judgments about what has happened, given the current situation, based on the assumption that systems become more disorderly over time. The presence of chocolate pudding is reasonably thought to indicate the past existence of separate quantities of milk, cocoa, and gelatin, but it doesn't work the other way: Seeing some quantities of milk, cocoa, and gelatin does not lead us to infer the past existence of some pudding from which their separate existences were derived. Making pudding is a one-way affair, and the kind of one-wayness we are talking about here seems to coincide nicely with the direction of time. Is thermodynamic asymmetry, then, just the same thing as temporal asymmetry?

The coincidence of the thermodynamic 'direction' and the direction of time is striking, but if the direction of time is real and unchanging, they can't just be one and the same, because the thermodynamic law in question is merely probabilistic. It is not impossible for the overall orderliness of a system to increase—just highly improbable. Take ink and water again. If you drop some ink into a glass of water, it will tend to diffuse itself more or less evenly in the water; any other result would be highly unexpected (see figure 5.1).

There is no force or rule that prevents every molecule of ink diffused in water from collecting in one spot; it is just that it is preposterously unlikely that every molecule of ink will happen to be

---

* We know that, just after the Big Bang, the universe was in a mostly homogenous state. Just why the universe was ever in this orderly (and thus unlikely) state is one of the primary questions in physical cosmology. If disorderliness is very much more likely than orderliness, then wouldn't it be very unlikely that the universe should have *ever* found itself in such a simple state? The answer to this requires a theory as to the origin of the universe, which is an issue I discuss in chapter 8.

Figure 5.1. Ink diffusing in water. This process could spontaneously occur in reverse without violating any natural laws.

randomly heading in that direction at the same time. Further, even if something so unlikely should happen, we would not want to say that the liquid was moving backward in time. It would just be a case of a process working in an improbable manner.

Now suppose that, by wild happenstance, all processes similarly trended in an orderly manner, such that the whole universe, over time, became a more orderly place.* Would this constitute a reversal of time itself? It would only if we give up on the idea of a fixed direction of time. Giving up on this idea, it seems to me, is just to give up on a real direction of time. It would also seem arbitrary to say that time is reversed if the whole universe becomes more orderly, but it isn't reversed for any particular subsystem that becomes more orderly. Identifying time with the thermodynamic arrow thus would make the direction of time a contingent, local, and potentially temporary phenomenon.

THE CAUSAL ANALYSIS

Finally, there is the causal arrow, which, like time, is asymmetric: According to the ordinary notion of cause, causes always precede

---

* Notice that I said "*became* a more orderly place," which presumes a direction of time. This is a consequence of lacking a convenient language that doesn't include dynamic temporal ideas as part of its structure.

their effects. Leibniz was the first to formally propose that time's arrow is to be understood in terms of causation. Leibniz, the reader may recall, was a temporal relationist. He described time simply as "the order of non-contemporaneous things." Like Aristotle, he felt that time was not something real in itself, but was rather our way of representing or measuring change. Here is Leibniz explaining how thinking in terms of time *order* is itself merely our way of representing causal dependency:

- *If many states of things are assumed to exist, none of which involves its opposite, they are said to exist simultaneously.* Thus we deny that the events of last year are simultaneous with those of this year, for they involve opposite states of the same thing.
- *If one of two states that are not simultaneous involves the reason for the other, the former is held to be earlier, the latter to be later.* My earlier state involves the reason for the existence of my later state. And since, because of the connection of all things, my earlier state involves the earlier state of the other things as well, it also involves the reason for the later state of those other things, so that my earlier state is in fact earlier than their later state too. And therefore *whatever exists is either simultaneous with, earlier than, or later than some other given existent.*
- *Time is the order of existence of those things that are not simultaneous.* Thus it is the universal order of changes, when the specific kind of change is not taken into account. (Loemker, trans.)

According to the relationist Leibniz, what is real is the causal relation; the earlier-later relation is just our way of understanding or representing the causal direction. So for Leibniz, an event A is 'earlier than' some other event B, in that A is one of the causes of B (or coexists with one of the causes of B). This definition would account for an inherent temporal direction by identifying an objective, non-probabilistic basis for the difference between earlier and later. Further, the fact that causes always precede their effects would easily explain psychological asymmetry: On this account, memories pertain to the 'past' just because they are caused by the events of which they are memories.

Unfortunately, although the causal analysis of temporal asymmetry may look attractive in some ways, it appears to presuppose the very concept that it aims to explain. The problem is that of coming up with a definition of causation that does not itself rely on the concept of temporal asymmetry. It would seem natural to define a cause as *an event that brings about another event,* or, alternatively, as *an event that precedes another event according to natural laws.* But each of these employs the notion of temporal precedence in defining the concept we want to use to analyze temporal precedence. We can't solve this problem by defining a cause either as *an event necessitating another,* or as *an event increasing the probability of another* (i.e., A causes B if events of type B always come after events of type A, or if events of type B typically come after events of type A). Remember—as we noted in discussing the thermodynamic arrow—that no sequence of events is truly irreversible according to the laws of nature. Thus any 'effect,' consistent with the laws of nature, could, in principle, be equally followed by what we would be calling the 'cause.' In an even more unlikely yet possible scenario, 'effects' of that sort might just by sheer luck *always* be followed by 'causes' of that sort. So the fact that A causally necessitates B doesn't by itself give us a direction: Once again, we would need to add that A *precedes* B, which assumes the temporal asymmetry that the causal analysis seeks to explain.*)

---

* There are other ways people have tried to define 'cause' without referencing time, but these definitions all seem to have insuperable difficulties. I refer the reader to Jonathan Schaffer's article on the metaphysics of causation in the online *Stanford Encyclopedia of Philosophy* for an overview of the different candidate analyses of causation.

One account of causation views the causal relation as a primitive, unanalyzable notion. Primitive concepts are ones that are so fundamental to thought and reality that they cannot be defined in any terms other than themselves. (Other concepts that have been thought candidates for primitiveness include 'existence' and 'truth.') In other words, any definition of causation would be unavoidably circular. This might be thought to solve the above problem, in that our problem with the causal analysis of temporal direction is that any attempt to analyze causation seems to involve time; if causation is primitive, the problem is solved by fiat, as it were.

One big problem with holding causation to be primitive, however, is that this would seem to make knowledge of causal relations impossible. As David Hume pointed out, we do not experience causal relations over and above experiencing the related events themselves: All we ever actually experience are associations of events. It is the experience of these sequences of events that leads us to the idea that the events are causally related. But if causation is primitive, then it is not to be understood in terms of patterns of events. So no knowledge of events and how some type of event tends to follow another type could constitute knowledge of causal relations. Consequently, such knowledge would appear to be impossible on the theory that causation is not in any way to be analyzed in terms of anything else.

Further, the idea that the direction of time is fixed by the causal arrow would be undermined by the possibility of traveling to an earlier time. As we shall see in the next chapter, time travel is neither logically nor physically out of the question. The possibility of time travel to the past would tend to further undermine the identification of the temporal and causal arrow, because later events could then cause earlier ones.

DOES THE SO-CALLED DIRECTION OF TIME EVEN MAKE A DIFFERENCE?

Let's return, for a moment, to the diffusion of ink in water. We have talked about how it is possible, if highly unlikely, for such a process to reverse itself and turn into a process of concentration, wherein the ink molecules happen to gather themselves together. Note how the very description of the thermodynamic probabilities presumes a direction to time. The ink, we say, diffuses. If you look at the same process backward, you would be looking at a process of concentration. But why call that the "backward" direction? What entitles us to speak in terms of the movement toward diffusion and disorder as the "forward" direction and the movement toward concentration and reverse entropy as the "backward" direction? The statistical tendency *toward* (as opposed to *away from*) disorder is only the norm if we have assumed that time has a direction, and that processes tend to become more disorganized in that direction. Even calling a process "diffusion" rather than "collection" betrays the fact we have already established a direction for the process. We call it "diffusion" because we remember the ink being more concentrated and anticipate it being less concentrated; that is the psychological arrow, but we have not determined the nature of the relationship between the psychological arrow and the arrow of time. The thermodynamic and causal arrows, as we have seen, can't function non-circularly as accounts of time's direction. It is totally arbitrary for us to just stipulate that the typical entropic or causal direction is the same as the time direction.

Clearly, psychology is the main reason why we tend to assign a particular direction to time. We remember only in one direction ("backward") and anticipate in the other ("forward"). Why do things work this way? Why do we remember the past and not the future? At one point, Stephen Hawking proposed that memory and entropy are linked. In forming a memory, we reconfigure our neurons. This creates a local increase in order (within parts of our brain responsible for memory), but only at the expense of a slight expenditure of energy, a dissipation of bodily heat, and an overall entropy increase. Therefore, on this account, the formation of memories is related to the larger thermodynamic trend. Our brains getting themselves into better order happens within the context of the trend toward overall heat dissipation. In a universe where systems necessarily decrease in entropy, our brains couldn't be getting themselves into better order. According to this theory, then, the psychological arrow is dependent on the entropic arrow—and is thus just as contingent as the entropic arrow. Suppose we lived in a reverse-entropy universe, in which entropy was decreasing (i.e., things were becoming more ordered). In that universe—according to the theory linking the entropic and psychological arrows—we would remember the 'future' and anticipate the 'past.' *But the past, in that universe, would be just like our future, and vice-versa.* What difference would that make? (See figure 5.2.)

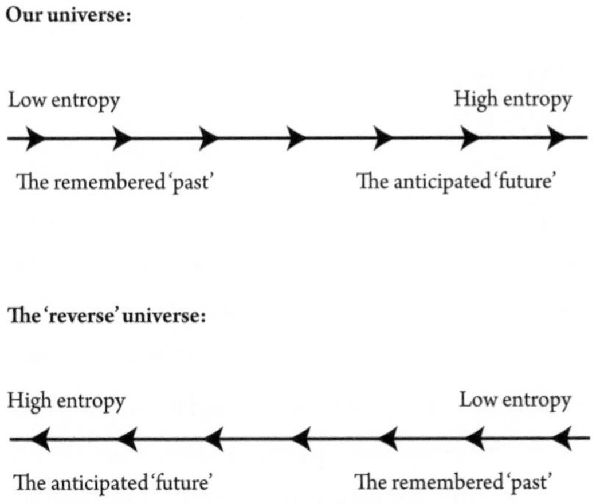

Our universe:

The 'reverse' universe:

Figure 5.2. What is the difference between these two scenarios?

The notion of a direction doesn't really seem to be adding anything to the story of how and why things appear the way they do to us. There is no more reason to call the psychological past and future, respectively, the "backward" or "forward" direction than there is in the case of the thermodynamic or causal arrows. As it is with what time it is *now*, the *direction* of time means something to us, but it appears to mean nothing to the universe.

McTaggart suspected all along that, in addition to the notion of the moving present, the notion of an earlier-later direction of time is based on a confusion and only has subjective significance. Thus he proposed that events are ordered only as a **C-series**, an array of timeless moments that are not only static but also lack a direction. The English alphabet is an example of a C-series: It so happens that we are used to reciting it from A to Z, but the list of letters itself has no inherent direction from A to Z. The above considerations about time's arrow suggest that the alleged directionality of time is, similarly, based on a mere convention as it arises from human psychology.

QUANTUM ENTANGLEMENT AND THE DIRECTION OF CAUSATION

This conclusion is further suggested by the behavior of certain microphysical processes. In the quantum realm, or the realm of the very, very small, scientists have observed phenomena that call our presumptions about the nature of physical law into question. For example, the setting on an instrument used to measure the behavior of spatially separated photons appears to affect the state they are in (their direction of spin, for example): The state the photons are in is correlated in a way that would seem impossible unless they are able to instantaneously communicate with each other, which itself seems impossible. This **nonlocality** or, in Einstein's dismissive words, "spooky action-at-a-distance," conflicts fundamentally with standard relativistic physics and so appears to undermine it. Yet this sort of phenomenon is undoubtedly real. The issue is how to interpret it. Prevailing interpretations include allowing nonlocality and thereby rejecting classical and relativistic physics altogether.*

Huw Price, Australian philosopher and founder of the University of Sydney's Centre for Time, is puzzled by this interpretation of such quantum correlations. He thinks that this doesn't take the static theory's block universe perspective seriously. In particular, if you take the block universe proposal seriously (as physicists do), then why presume that causation only works in the so-called forward direction? That is to say, if we truly live in the timeless reality described by physics, then why can't the future affect the past just as the past affects the future? What Price proposes is that what we are seeing in the photon experiments is **retrocausation**: The *later* measurements are causally influencing the *earlier* properties of the particles! That influence would explain their observed correlation without instantaneous nonlocal influence, without our having to revise classical and relativistic physics.

The same goes for familiar claims in quantum physics as to the **indeterminacy** of the states of elemental particles before they are measured. Experiments similar to the one suggesting nonlocality have typically been taken to imply that subatomic particles literally remain in indeterminate states until observed. Such claims culminated in the famous "Schrödinger's Cat" thought experiment. Austrian physicist Erwin Schrödinger thought this indeterminacy was absurd. He explained how one could put a cat into a sealed box

---

* Other, even more ghastly suggestions include the existence of an infinity of simultaneously existing alternate universes. I address that idea a bit more in the last chapter.

with some poison, such that the release of the poison is dependent on the state of a subatomic particle. If the indeterminacy interpretation of these quantum phenomena is correct, we would have to conclude that the cat is both alive and dead until the box is opened; and only at the moment the box is opened does the cat enter into a determinate state of being alive or being dead. His point was that there had to be something wrong with any interpretation that leads to this conclusion.

The proposal that causation can work 'both ways,' so to speak, has the advantage of allowing us to dispense with the "conceptual horrors" (as Price calls them) of nonlocality and indeterminacy. No more spooky action at a distance, and no more cats that are both alive and dead.

Why does the notion of backward causation, or retrocausation, seem so strange? Only, Price explains, because we are so used to seeing the thermodynamic arrow point in one direction when it comes to the world of our ordinary experience, the macrophysical world. He argues that, if we had approached the quantum state experiments from the standpoint of a timeless universe in the first place, we would have seen these correlations as straightforward evidence of retrocausation. It is only a deep but inherently unsupported prejudice against the so-called later explaining the so-called earlier that gets in the way of an otherwise simple explanation of these quantum correlations.

Price supports the idea, described earlier in this chapter, that the arrow of causation, as well as the arrow of time, is a projection onto an inherently time-symmetrical world, deriving from our manner of processing that reality. Causation, he asserts, is a mere human projection, related to the statistical thermodynamic asymmetry of the macrophysical world—the world pertinent to human psychology. He explicitly compares the notion of one-way causal relationships to sensory properties like color: The ascription of causal powers to events is the result of psychological projection, just as the ascription of color to objects is. The feature of our experience that results in our ascribing color to objects, of course, is just the visual sensation itself. "What feature of our perspective could it be," Price asks,

> that manifests itself in the cause-effect distinction? The most plausible answer is that we acquire the notion of causation in virtue of our experience as agents. Roughly, to think of A as a cause for B is to think of A as a potential means for achieving or bringing about B.... The origins of causal asymmetry thus lie in our experience of doing one thing to achieve another—in the fact that in the circumstances in which this is possible, we cannot reverse the order of things, bringing about the second state of affairs in order to achieve the first. This gives us the causal arrow, the distinction between cause and effect. The alignment of this arrow with the temporal arrow then follows from the fact that it is normally impossible to achieve an *earlier* end by bringing about a *later* means.

Whereas, in fact, if he is right, we *can* bring about a change to the earlier state of subatomic particles by measuring their state at a later time. It is just that we did not evolve with this sort of causation as a part of our experience, and thus as relevant to decision-making.

Price concludes that the true "view from nowhen"—a perspective treating the space-time continuum like the timeless block it is—would include neither an inherent time direction nor an objective, one-way causal arrow.

Price's interpretation of experimental observations in quantum physics remains an interpretation, and a controversial one at that; again, however, it has the big advantages of (a) elegantly allowing us to dispense with the counterintuitive notions of nonlocality and indeterminacy, and (b) taking seriously the true symmetrical nature of time and the inherent subjectivity of time's arrow. In light of everything we have seen pertaining to the philosophy and physics of time, it is a compelling story.

WORKS CITED IN THIS CHAPTER

Hawking, Stephen. *A Brief History of Time* (New York: Bantam, 1988).

Leibniz, G. W. "Metaphysical Foundations of Mathematics," in *Philosophical Papers and Letters*, trans. by Leroy Loemker (Chicago: University of Chicago Press, 1956).

McTaggart, J. M. E. "The Unreality of Time," *Mind* 17 (1908), 457–474.

Price, Huw. *Time's Arrow and Archimedes' Point* (Oxford, UK: Oxford University Press, 1997).

Schaffer, Jonathan. "The Metaphysics of Causation," in *The Stanford Encyclopedia of Philosophy*, ed. by Edward N. Zalta, http://plato.stanford.edu/entries/causation-metaphysics/.

Other works of relevance to the issues in this chapter

Carroll, Sean. *From Eternity to Here* (Oxford, UK: Dutton, 2010).

Van Fraassen, Bas. *An Introduction to the Philosophy of Time and Space* (New York: Random House, 1970).

Chapter 6

Is Time Travel Possible?

The static theory of time treats temporal location a lot like spatial location. In accordance with contemporary physics, space-time is treated in this theory as an unchanging four-dimensional block. Each temporal slice of the continuum is just as real as any other, just as any part of space is just as real as any other. Some might suggest that this means that travel to other parts of time should be possible, at least in theory, just as travel to other parts of space is possible. Thus an investigation of the possibility of time travel is at the same time an investigation of the implications of the static theory of time.

FICTIONAL TIME TRAVEL

There are numerous examples of fantastical descriptions of time travel in literature and film from the last century or two. From the H. G. Wells book *The Time Machine* to Terry Gilliam's movie *Twelve Monkeys*, many exciting tales have been spun involving a leap of some sort from a present time to a past or future time. The time traveler in *The Time Machine* builds a machine (the operating principles of which are unclear) that takes him to a far future state of the Earth. The protagonist of *Twelve Monkeys* travels (once again, technical details are fuzzy) to a past state of the Earth, where he gets involved in events that turn out to have a lot to do with conditions in the era from which he came.

As much as narratives like these capture the imagination of readers and moviegoers, they typically do not take into account a serious consideration of either the logic or physics of time. What sort of travel do these storytellers have in mind? Colloquially, what we usually mean by "travel" is travel *through* space (say, from New York City to Paris) *over* time. Although it does take time to get there, the destination is defined by its *spatial*, not its *temporal*, location. The very notion of time travel treats temporal locations like spatial ones. Although this metaphor might work under the suspension of disbelief we allow ourselves in enjoying a work of fiction, it doesn't just go without saying that time travel is a coherent notion, much less a plausible one. However, now that we have discussed relativistic physics and the distinction between the A-series and B-series of events, we are in a much better position to examine the possibility and implications of time travel.

Right off the bat, just based on what we have covered so far in this book, there are some reasons to be optimistic about the possibility of some form of time travel. We have learned that the most plausible model of space-time is the block model, which includes a timelessly existing span of events. If all past, present, and future events timelessly coexist, then at least there is a potential destination for the time traveler.*

Further, the special theory of relativity describes different reference frames defined by relative velocities, each with its own perspective as to which events are simultaneous with others. One consequence is that what constitutes the present is, in a quite

---

* The presentist, by contrast, treats past and future events as nonexistent. There is no place for the time traveler to go.

substantive sense, a matter of perspective. This seems to open the door to the notion of shifting from one 'present' to another via changes in one's velocity (and thus changes in one's reference frame). As we shall see shortly, the general theory of relativity suggests some possible tricks for jumping from one space-time location to another via the manipulation of space-time itself.

In order to examine the possibility of time travel in reality, as opposed to fiction, we need to determine first if it is even a logically coherent notion. If it is, then it makes sense to ask whether the laws of nature as we understand them might allow it. If they do allow it, then we can discuss whether any sort of time travel is feasible, given the available resources.

IS TIME TRAVEL LOGICALLY POSSIBLE?

When we ask whether time travel is possible, we can mean any of three different things, corresponding to three senses of 'possibility': logical possibility, physical possibility, and practical possibility. **Logical possibility** refers to whether the very notion of something implies a logical contradiction. For example, constructing a round square is not logically possible, as it would involve an item with contradictory properties. An event is **physically possible** if its occurrence does not violate any natural laws. Logical possibility trumps physical possibility: Events can be logically possible without being physically possible. Think of a round square again: If we know a round square is logically impossible, we don't need to look into whether physics allows such a thing. Conversely, events that are physically possible cannot fail to be logically possible: For example, though it may run contrary to the laws of nature for an unsupported pencil to hover in the air near the surface of the earth, this phenomenon implies no logical contradiction. Finally, there is **practical possibility**, or feasibility: Practical possibility simply refers to what one can actually do given one's available resources. There are many accomplishments that would not violate any laws of logic or nature, but for which the energy or materials are simply not available.

Because logical possibility trumps the other kinds, it makes sense to start with the question as to whether time travel would involve any logical contradictions. Time travel, at least as it is often represented in fiction, does seem to allow for logical contradiction or paradox. The most often-cited paradox is the **grandfather paradox**. Take the usual science fiction variety of time travel, involving a vehicle of some sort that takes you into the past. If this sort of time travel is possible, then it should be possible to go back in time and kill your own grandfather as a young boy, thereby preventing your own birth. But, of course, that would mean that you would not exist and be able to go back in time and kill your own grandfather, etc.

The grandfather paradox is, indeed, a serious challenge to the logical possibility of time travel. Furthermore, the static theory of time entails that past and future events are unalterable, because what we call past and future events are really just timelessly existing earlier and later events. You cannot change unchanging events; if time travel suggests you can, then so much the worse for time travel. If these concerns rule out time travel on logical grounds, then we don't even need to do a scientific study of the possibility of time travel; we would already know that it can't happen.

In fact, time travel and the static theory can be logically consistent. The static theory states, essentially, that what is done is done. The past BE just the way that it BE, and there is no changing it. This does not necessarily rule out time travel, however, as this fact does not rule out the possibility that the past includes the actions of time travelers: Any past actions of time travelers simply BE timelessly included in the history of the world. This means that, even if people had traveled to their pasts, they *couldn't* have killed their own grandparents. Why? Because they *didn't*. This sounds odd, because, if we grant the availability of a time machine and a pistol, why would one not be able to carry out the action in question? Philosopher David Lewis argued that this apparent oddness is just due to the fact that the issues relevant to deciding the truth of a claim that one "can" do something are context-dependent. It may be perfectly true that, given the necessary resources and opportunity, you "can" murder another person; however, given that you exist at a later time, it is not the case that you "can" (in a different sense) kill your grandfather as a young boy. Even if you really went back in time to do it, something must have intervened in the attempt (a misfire, perhaps, or change of heart), because your assassination attempt has already happened and was, evidently, unsuccessful. Even if science fiction time travel is possible, the fact is that your grandfather BE alive, and you BE his grandchild. You can no more change such facts, as physicist Brian Greene puts it, than you can change the value of *pi*.

So the grandfather paradox, combined with the fixity of the past and present in the static theory, does not rule out the possibility of time travel; it just rules out inconsistent stories about how the world is throughout time. As long as we understand any time travel events to be timelessly included in the history of the world, and thus as part of the fixed continuum of events, time travel need not give rise to paradox. However, the logical possibility of time travel does not mean it is physically possible. Does nature, at least as we understand it, allow for time travel?

IS TIME TRAVEL PHYSICALLY POSSIBLE?

Again, the standard science fiction scenario is the idea of a machine that transports you backward (or forward) through time. The more common backward-traveling scenario is problematic right off the bat because of the "double-occupancy problem," cited by Robin Le Poidevin and others. Suppose you press the button and the machine starts traveling back in time. The machine traveling back in time must, shortly afterward, be occupying about the same space that the machine was occupying (while it was traveling forward in time) one microsecond before you pressed the button. But two solid items cannot occupy the same space at the same time.*

More promising are some other more scientifically grounded proposals about time travel relating its possibility closely to Einstein's general theory of relativity (GTR), which states that space-time can itself be distorted in response to the presence or activity of matter. This effect has led some to suggest how time travel might be physically possible in relativistic space-time. Late in his career, at Princeton, Einstein befriended the great mathematician Kurt Gödel. Their conversations led Gödel to work out the mathematics for a GTR universe in which all matter was rotating; in such a universe, space-time would be so distorted that some light cones would tilt far enough over to allow travel into one's own past.

Physicist Kip Thorne and others have, more recently, argued that GTR also predicts, in the presence of enough mass or energy, the formation of "wormholes"—tunnel-like distortions of space-time

---

* If you change the story and make it a machine that disappears completely and reappears in some other time, then you have a metaphysical problem about personhood and continuity: Namely, can the human occupant of the machine really be the same person after experiencing a complete discontinuity like that? Or would such a machine just have the effect of creating a duplicate of its passenger in the past? The latter would give you backward causation, but it would not constitute *transportation*.

that would allow passage from one area of space to another, and which could be used, in theory, to travel to one's own past.

Despite being endorsed by serious thinkers, proposals like these regarding the theoretical possibility of time travel are matters of ongoing scientific controversy. Certainly, even if such travel is physically possible, no sort of time travel relying on exotic space-time configurations looks practically possible: The energy needed to distort space-time in the ways described would be cosmic in scale. Nor is it uncontroversial that these proposals really do describe physically possible time travel methods. Einstein suggested that Gödel's model might not take into account other physical laws that would prevent time travel even in a rotating universe; others have argued that wormholes are inherently unstable, or could not be used for time travel for other physical reasons. So there is no consensus that backward time travel is physically possible. Tim Maudlin describes the speculations of Gödel and Thorne as exercises in coming up with space-time configurations that are mathematically consistent with GTR, but which would or could never actually take form. What is the relevance, he asks, of such alleged possibilities to our conception of time?

Should claims about the legitimate physical possibility of time travel hold up, the metaphysical implications would be very interesting. The actual, physical possibility of time travel, even if it isn't practically possible, would mean yet another blow to the dynamic theory of time. According to the dynamic theory, there is one path through time, with one past, one present, and one future; any event, and every moment, has the property of pastness, presence, or futurity. If one could, even if only in principle, travel to the past, then past events would be potentially present for the traveler, present events future, and so on. So such properties could not really be properties of events; what is past, present, and future would have to be a subjective matter. Indeed, Gödel rejected the dynamic theory of time for this reason.

We would also have to rethink causation and the arrow of time. If time travel to the past is consistent with the laws of nature, then past events could be caused by future ones; this would lend further credence to the notion, put forth by Huw Price and others, that the one-way directionality of time is not an inherent aspect of nature.

A QUESTION ABOUT TIME TRAVEL ASYMMETRY

So far we have only talked about travel into the past. If passage is real, then we travel into the future, so to speak, all the time. We also know, thanks to Einstein's special theory of relativity, that it is really pretty easy to 'slide' into the future simply by accelerating oneself around for a bit before returning to one's point of origin. It is a familiar consequence of the theory of relativity that someone departing from Earth on a fast spaceship, after accelerating away sharply and later returning, will find a difference between his or her clock and clocks back home. Imagine a pair of twins, one of whom departs for just such a journey. Upon returning, she will find that time has 'passed' more slowly for her than for her twin back on Earth.

One moderately helpful way to understand this bizarre phenomenon is by thinking of each of us as having a path through space-time. Each of us traverses the same amount of space-time over time, but not necessarily in the same way. The more distance one traverses in the spatial dimension of space-time, the less of the temporal dimension one will have traveled, and vice-versa (within certain limits). The fact that the one twin takes this trip means that she winds up measuring the passage of time from the standpoint of two different frames of reference during her trip, instead of one. The respective paths through space-time for our separated twins could be represented as shown in figure 6.1.

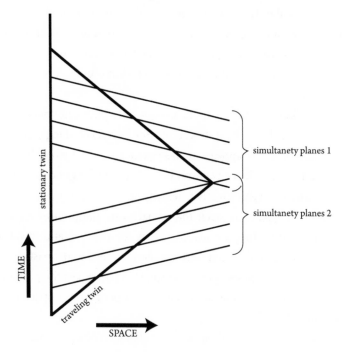

Figure 6.1. Each plane of simultaneity represents what spatially separated events the traveling twin would measure as simultaneous with each other.

Putting it roughly (and from the stationary twin's perspective only), the traveling twin's nonlinear path through space-time takes her through more of the spatial dimension but less of the temporal dimension than her twin.* She ages less while covering more distance.

---

* Also note that the space and time axes are labeled on behalf of the stationary twin only; the traveling twin looks at space and time differently by virtue of her different reference frame(s), which of course is essential to the whole idea of relativity. Figure adapted, courtesy of Wikipedia user Acdx, based on file: Twin paradox Minkowski diagram.png. Accessed Oct. 22, 2012, at http://en.wikipedia.org/w/index.php?title=File:Twin_Paradox_MinkowskI_Diagram.svg.

This effect of acceleration has been confirmed many times over by experiments. A very accurate atomic clock on board an ordinary passenger jet will noticeably diverge from its twin back on the ground after a few flights, precisely to the degree predicted by Einstein. This makes anyone who goes on a plane trip a time traveler into the future, even if the time difference upon return is only a tiny fraction of a second.

Slides into the future, then, are not only physically but practically possible. In contrast, travel into the past, even if logically and physically possible, appears to be practically impossible. In considering this fact, Barry Dainton has asked a really good question: If the static theory of time is true, why should a forward slide into the future be so much easier than a backward jump into the past? If the past and future are just as real as the present, and travel to the future is possible (even routine), why is travel to the past either wildly difficult or physically impossible? What makes the past and future so different if they are really just parts of a static space-time continuum?

Despite the necessarily speculative character of any discussion of time travel, it is a useful area of examination because it gets us to ask some good questions, like Dainton's question about the asymmetrical nature of time travel. It is a question that demands we maintain some skepticism that the static theory fully captures the nature of time—at least until we understand more.

There is yet another kind of alleged asymmetry between past and future that accounts for a common claim about the inadequacy of the static theory: Though we may all agree that what is done is done, human freedom requires an open future of genuine alternative, undetermined possibilities. The static theory appears to be incompatible with freedom, because it does not allow changes to any part of the universe's timeline, including the future. Consequently, if we are free, the static theory must be wrong. Whether this line of attack really does give us a reason to think that the static theory is wrong is the subject of the next chapter.

WORKS CITED IN THIS CHAPTER

Dainton, Barry. *Time and Space* (Montreal: McGill-Queen's University Press, 2002).

Le Poidevin, Robin. "The Cheshire Cat Problem and Other Spatial Obstacles to Backward Time Travel." *Monist* 88 (2005), 336–352.

Lewis, David. "The Paradoxes of Time Travel," *American Philosophical Quarterly* 13 (1976), 145–152.

Maudlin, Tim. *The Metaphysics within Physics* (Oxford, UK: Oxford University Press, 2007).

Other works of relevance to the issues in this chapter

Callender, Craig. *Introducing Time* (New York: Totem Books, 1997).

Carruth, Shane. *Primer* [movie] (2004).

Earman, John. *Bangs, Crunches, Whimpers, and Shrieks* (Oxford, UK: Oxford University Press, 1995).

Epstein, Lewis Carroll. *Relativity Visualized* (San Francisco: Insight Press, 1985).

Falk, Dan. *In Search of Time* (New York: St. Martin's Press, 2008).

Gilliam, Terry. *Twelve Monkeys* [movie] (1995).

Hawking, Stephen. *A Brief History of Time* (New York: Bantam, 1988).

Chapter 7

Time and Freedom

Is the future already written? According to the static theory of time, all so-called future events timelessly BE, distributed over the temporal dimension of space-time. Because static theory seems to suggest that all facts just are what they are, with no distinction between the future and the 'settled' past, it might be taken to imply that we cannot change the future: If there is only one way for things to go, then we can't choose otherwise than we in fact do. How can we be free, if we can't choose otherwise than we do? Is the impossibility of free will really a consequence of the static theory of time? (And, if so, would this consequence be a reason to reject that theory?)

ARISTOTLE AND THE SEA BATTLE

Fatalism is the view that human decision-making is irrelevant because the future involves no genuine alternative possibilities. In the philosophical context, this does not refer to active intervention on the part of any mystical or supernatural forces, but rather to certain logical, metaphysical, or theological problems with the notion of an open future. Fatalism conflicts with a deeply held intuition that we can sometimes choose between genuine alternatives, and that some events are in some respect contingent on our decisions.

The static theory of time would be much harder to accept if it were to require us to embrace fatalism. Defenders of the dynamic theory of time sometimes argue against the static theory precisely on this basis: Some claim that the static theory can't be right because human beings have free will, and free will requires an open future. Proponents of the dynamic theory correctly point out that, on the static theory of time, the future is written, down to the last detail: Whatever 'will' happen timelessly *happens*. At first glance, this appears to rule out a future of different, real possibilities that can be resolved one way or the other by human decision-making.

To put it more formally, the problem is that the static theory appears to lend support to **logical fatalism**, which we might also simply call generic philosophical fatalism. This version of fatalism dates back to Aristotle and his contemporaries, and refers to a simple logical conundrum about future possibilities. Aristotle's extensive work on logic was one of his many contributions to philosophy. In one text, he considers a puzzle about whether there are any genuine possible, yet non-actual, events. Take the following two propositions:

1. There will be a sea battle tomorrow.
2. There will not be a sea battle tomorrow.

Normally, we would like to say that each of these propositions about tomorrow's battle expresses a genuine possibility, up until the day arrives and the weather either permits the battle or forces the respective navies to delay. Upon reflection, however, this is not so clear-cut. An intuitive and widely accepted logical rule dictates that every meaningful, factual proposition must be either true or false. This is called the **principle of bivalence**. Another widely accepted rule states that no proposition and its negation can both be true at the same time; this is the **principle of non-contradiction**. If we accept both these rules, we must concede that, the day before the expected battle, one of the propositions about the sea battle's occurrence is now true and the other is false (even if we don't know which). Of course, the same will be true for any other proposition about the future. The fatalist concludes from this that any event in the future that turns out not to come to pass was never really a possibility in the first place, no matter how plausible or even likely it seemed; and any event that does occur could not possibly have failed to occur. This would mean that nothing that will happen in the future could happen otherwise. Logical fatalism does not refer to any *causal* necessity attaching to future events, in that the issue is not that the laws of nature dictate, given the current situation, what must happen next; the issue is whether logic permits two contradictory propositions about the future to each be genuinely possible. The key consequence of logical fatalism pertains to human freedom of choice and action. It would appear that, if logical fatalism is true, then we cannot be free: *If it is true now that any given future event definitely will (or will not) happen, then there is no way for us to change the future course of events.*

Aristotle rejected this result. His answer was that propositions about the future constitute an exception to the principle of bivalence. Contrary propositions about the future, he explained, describe only potential situations, not actual ones. It is true that one or the other of two contrary propositions about the future must be true at some point, but neither must be true now.

Aristotle's picture coincides with what has to be regarded as the standard, non-philosophical view that most people have about human freedom. According to this standpoint, human choices play a role in shaping as-yet non-actualized future events. The future is not established or actualized until choices are made and their consequences play out. Human choices are based on reasons, but our choices can go either way until we decide which reasons are most important.

There are a few possible objections to Aristotle's solution.* The biggest problem has to do with its presumptions about time. Obviously, Aristotle's answer presumes the dynamic theory of time. He assumes a difference between the past and the future, in that past facts are settled, and future facts are merely potential. The merely potential status of future facts changes with human choices and the passage of time.

This answer is undermined by what we have said in favor of the static theory of time. On the static theory, there is no actual versus potential distinction between so-called past and so-called future events. Everything happens timelessly, in different spatial and/or temporal slices of space-time. Every event (as we know from relativity) is both past and future in some frame of reference, and no such perspective is privileged. Every event BE just the way that it BE. So, the fatalist might claim, the static theory of time entails no open future and no free will. This position we shall call **metaphysical fatalism**, which supports the key claim of logical fatalism: namely, that all factual statements are either true or false—including ones about 'the future' (i.e., what counts as the future from one's perspective at the

---

* One technical issue with Aristotle's answer is that it seems to make accurate prediction impossible. We would like to say that if one predicts a sea battle, and it comes to pass, then one spoke truly about the future battle at the time the prediction was made. But Aristotle's solution, because it describes future events as mere contingencies, means that, though you can truly say that a given future event is possible, you can't speak truly or falsely about its actually coming to be in advance of its coming to be. Yet we tend to think one can be congratulated upon making an accurate prediction—indeed, betting on sports is predicated on this idea. John MacFarlane replies that the truth value of a prediction is context-dependent: Whether it should be thought of as having a truth value depends on the context in which it is being assessed. He claims, in other words, that a statement like "There will be a sea battle tomorrow" is neither true nor false before the battle, but it can be retroactively assessed for accuracy after the fact. It is not clear who is right about this, but it is a moot point if there are other, compelling reasons to reject the dynamic theory of time (as there seem to be).

time of utterance). Metaphysical fatalism's embrace of the principle of bivalence derives from the static theory's metaphysical position on the status of events in time.

A connection between the static theory and fatalism would be embarrassing for the theory. It would make a hash of the usual distinction between necessary and contingent events by suggesting that only one sequence of events is really a possibility at any moment. Fatalism suggests that human influence on the sequence of events is not possible: Whatever we do, later events will be what they will be (because they BE what they BE). Yet surely there is some sense in which our choices can influence the course of events; history has played out a certain way, but surely there is the thought that things could have gone differently. Unlike truths of mathematics, facts about the course of events are not necessary truths. Can the static theory account for the contingency of events?

As we have seen, the static theory is well supported in many other respects. According to the static theory, if the sea battle takes place on a given date, then it always does so, regardless of when we might be considering the truth of a proposition about the battle. Whether the battle takes place in the 'future,' 'present,' or 'past' is a matter of psychological projection based on one's perspective at the moment of considering that event. There are no merely 'potential' events in the sense in which Aristotle is using the term. The occurrence or nonoccurrence of the battle is not logically necessary in the sense that the contrary state of affairs would involve a logical contradiction, but the truth of the correct statement about the battle is assured nonetheless, no matter when it is uttered. Aristotle's answer stands or falls with the dynamic theory of time, and we have seen in chapter 4 that neither logic nor physics favors that theory. If the static theory is correct, then the application of the principle of bivalence to propositions about the future seems appropriate.

Happily, however, this does not mean that the static theorist is committed to fatalism. L. Nathan Oaklander agrees with the static theory of time, but denies that fatalism is a consequence of that theory. He agrees that the fact that a given event occurs, if true, is timelessly true. But this does not mean that the truth of any such fact cannot depend on a choice someone makes at some particular moment. Take the timeless fact that Napoleon signs the Concordat of 1801. Oaklander points out that there is nothing preventing the static theorist from acknowledging the timeless fact of this event, while attributing the existence of this fact to a decision made by Napoleon, shortly before the signing, to enter into an alliance with the Roman Catholic Church. The principle of bivalence requires that the statement "Napoleon signs the Concordat" is timelessly true, *but it does not require the cause of the truth of that statement to be in effect in advance of Napoleon's decision.* So we can agree that all statements about the future are either true or false, but we need not conclude that we have no control over future events. In this way the static theorist can have her cake and eat it too: She can uphold the timeless truth of all facts without requiring that every event is therefore inevitable.

Yet it must be noted that this answer does not clear the way for free will. Even if fatalism is not a problem, there still remains causal determinism, which represents a different sort of attack on the notion of our having control over events.

CAUSAL DETERMINISM

Aristotle's fellow Greek, Chrysippus (third century BCE), was highly regarded as a logician; he also focused on causation, freedom, and virtue. In relation to his interest in freedom, he discussed both fatalism and **determinism**, which is a doctrine that bears a superficial resemblance to fatalism, but involves a different issue altogether: causal, or natural, necessitation. Fatalism just makes the point that, if all statements about the future are timelessly either true or false, then all events 'must' therefore occur just as they do—regardless of whether they occur in accordance with natural laws. By contrast, determinism has to do with natural causation. Determinism states that all events have a cause, and thus all events are determined by a chain of causes going back uninterruptedly into the past. This appears incompatible with the notion of having genuine control over one's own choices. Human choices are just events in the world like any other, so they are determined by past circumstances—whether internal or external to the agent—just like any other events. To put it slightly differently, according to determinism, given the way the past is, plus the laws of nature, any event must (by the laws of nature) unfold the way that it does; because persons do not control either the past or the laws of nature, they cannot ultimately control their own choices, regardless of appearances. Note that this position about the causal determination of choice is not specific to either the dynamic or static theory of time, as each theory embraces the notion of causation and natural law.

The determinist standpoint is commonly linked to **scientific naturalism**, which is the position that all events are natural (as opposed to supernatural) and so explicable according to natural laws. French mathematician and astronomer Pierre-Simon Laplace, when asked by Napoleon about the role of God in his theory of celestial mechanics, famously answered, "Sire, I have no need of that hypothesis." If all events are natural, and all natural events take place in accordance with natural laws, then all events are determined by prior causes according to natural laws. Laplace illustrated this proposal through the thought experiment of a hypothetical omniscient predictor: a being who knows everything there is to know about the current state of the universe, as well as everything there is to know about the laws of nature. If naturalism is true, then an omniscient predictor would be able to predict the future with complete accuracy:

> We may regard the present state of the universe as the effect of its past and the cause of its future. An intellect that at a certain moment would know all forces that set nature in motion, and all positions of all items of which nature is composed, if this intellect were also vast enough to submit these data to analysis, it would embrace in a single formula the movements of the greatest bodies of the universe and those of the tiniest atom; for such an intellect nothing would be uncertain and the future just like the past would be present before its eyes.

Determinism thus follows from naturalism. Causal determinism is, arguably, an even greater challenge to free will than fatalism. Unlike fatalism, determinism is a challenge to free will even if Aristotle is right and the dynamic theory of time is correct: If all events proceed from prior causes, it wouldn't matter if future events really are non-actualized now in every sense of the term, because future events are predetermined by the current state of things.* Given the way things were in the past, plus the laws of nature, our own future choices are inevitable (see figure 7.1).

---

*I am ignoring quantum physics, which suggests that all unobserved events are properly described in terms of probability only. The sense in which indeterminacy rules in quantum physics is unlikely to have any relevance to freedom: Even if it could be shown that choices derive from some sort of random quantum fluctuation, this would do nothing to show that we have determinative control over our actions. More on this follows in chapter 8.

Dynamic Non-Determinism:

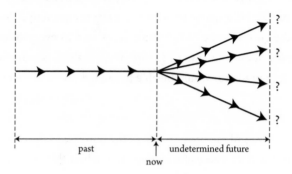

Dynamic Determinism:

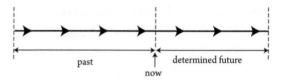

Static Fatalism:

Figure 7.1

Causal determinism has equal impact on the static theorist's account of free will. Oaklander, the reader may recall, is able to describe the compatibility of free will with the static theory by noting that there is nothing stopping us from explaining the timeless fact of an event by reference to a decision made at a time. This answer, however, says nothing about whether causal factors, ultimately beyond the individual's control, make that decision itself inevitable.

Laplace's omniscient predictor is just an illustration of an idea quite familiar to Chrysippus, who accepted causal determinism, as well as logical fatalism. Like Aristotle, his concern about the possibility of human freedom was based on his concern about virtue and responsibility. As with fatalism, determinism seems to imply that we are not free: If all events are determined by natural laws, and human choices are events in the world, then our choices themselves are ultimately determined by prior events over which we have no control. This situation seems to rule out the possibility of moral responsibility.

AN ANSWER TO BOTH FATALISM AND DETERMINISM: COMPATIBILISM

In response, Chrysippus proposed that free will, properly understood, is compatible with determinism. His answer, called **compatibilism**, remains the most popular philosophical answer to both determinism and fatalism. Chrysippus distinguished between an event that is causally determined to occur and one that is logically necessary. Logical necessity has to do with the converse of a proposition implying a logical contradiction. Whether we say an event is determined or fated, in neither case are we talking about this sort of logical necessitation. Even if they may be causally determined or fated to occur, human choices are not logically necessary. Chrysippus goes on to point out that we can distinguish between events that we help to bring about and events that are brought about by external causes alone. An action is free, he argues, if it stems from one's own intentions, as opposed to external causes. Intentional actions are free, in the sense that jumping off a cliff is a free act; whereas falling off a cliff after being pushed is not. Of course, the fact that one was going to have the intention to jump is causally pre-determined. Nevertheless, the proponent of compatibilism can distinguish between those events one can control in the sense that they occur *because one wants them to*, and those one does not control in that same sense. Chrysippus—like modern compatibilists—claims that the concept of moral responsibility applies to actions based on one's own intentions, even while accepting the prior determination of those very intentions.

Much later, Leibniz also embraced compatibilism in order to rescue freedom and moral responsibility from **theological fatalism**, a problem about freedom specific to theists. Theological fatalism is—or should be—of concern to believers in an omniscient deity. If God is omniscient, then God has perfect foreknowledge of everything we do. If God has perfect foreknowledge, then our future choices cannot be otherwise than they will be. How, then, can we be considered free? This concern shares aspects of both fatalism and determinism. It is a concern that can arise within the context of the dynamic theory of time; but if God exists and has perfect foreknowledge, then we don't have an open future in the sense of having genuine alternative possibilities: The way God foresees things happening is the way they will go.

Leibniz's answer to theological fatalism, much like Chrysippus' answer to determinism, was to distinguish between its being logically necessary that one do something and its being merely contingently true—and perfectly predictable—that one will do it. God's foreknowledge just means that He knows what we are going to do; Leibniz argued that this perfect predictability does not necessitate our actions, because it remains 'possible' in the logical sense of the word that we could do otherwise.

Unfortunately for Leibniz, compatibilism doesn't work very well in a theological context, if the point is to defend individual moral responsibility: According to his own understanding of things, God would be the one who intentionally created us with just such a nature that under the circumstances—also preordained by God—we would do exactly as predicted. How could God then turn around and hold us responsible for the consequences? For this reason, theists tend to reject compatibilism in favor of a stronger notion of free will.

This particular problem for the theist aside, compatibilism can also be employed against the metaphysical fatalism derived from the static theory of time: We can distinguish between timelessly existing events in which human intentions play a role and ones in which they do not. According to both logical and metaphysical fatalism, we cannot *change* the future. Yet we can *affect* the future; we can distinguish between events that are brought about by human intentional action and ones that are not.

The defender of free will typically replies that the compatibilist conception of freedom is too weak: It doesn't live up to the most natural understanding of what it takes to be free in one's choices. Kant famously characterized compatibilist freedom as a "wretched subterfuge" and called it "the freedom of the turnspit." Like many others, he felt that any conception of freedom that does not allow for a choice between two genuine, undetermined possibilities is not capturing what we commonly mean by freedom and fails to explain moral responsibility. Kant's fellow German, Arthur Schopenhauer, succinctly expressed the paradox within compatibilism with the aphorism "Man can do what he wills but he cannot will what he wills."

So much the worse for the common notion of freedom, reply determinists. A closer look at the non-compatibilist version of freedom reveals a fundamental conceptual confusion. Non-compatibilist, or **libertarian**, freedom states that a free act is a truly undetermined event that itself determines some aspect of an open future. This is simply incoherent. If one's actions are caused neither by one's intentions *nor* by some outside force, the only alternative is that they are random: Events that are uncaused are events that have occurred by chance. But actions due to chance are not free in anybody's view. The libertarian's notion of freedom says that free acts are *both* undetermined *and* under one's control, but these characteristics are incompatible. The problem with the libertarian's stronger notion of freedom is not only that it presupposes the self-contradictory ascription of A-series properties to events, but also that it is itself self-contradictory.*

The libertarian notion of a free action seems to be, as Oaklander puts it, "one that is caused by the person or self; a substantial agent whose spontaneous act of creation lies outside the sphere of scientific predictability and external causality." This free action must be strictly unpredictable, yet not uncaused in the sense in which a random or chance event is uncaused. The libertarian wants actions that are explicable in terms of the reasons the agent is acting on, yet non-determined in the Laplacian sense. But it is not at all clear that there is any space for an action to be both based on reasons and causally inexplicable. If an agent is acting on the basis of reasons, then it is a causal story, because there will be a causal story as to why the agent has come to have those reasons. If the agent is not acting on the basis of reasons, then his or her actions are not deliberate or intentional—so once again are not free in the libertarian sense.

So compatibilist freedom is the only coherent proposal. This is a contentious result for a lot of people, mainly because it is not at

---

* And that's why indeterminism owing to quantum randomness is irrelevant to this discussion. Decision-making stemming from random events cannot be characterized as under one's control.

all clear that compatibilism leaves room for a robust sense of moral responsibility. How can we hold a criminal responsible for his or her actions if we acknowledge—as the compatibilist must—that the intentions behind those actions were entirely a product of his or her genetics, history, and circumstances?

Indeed, punishment purely for the sake of moral retribution gets called into question by compatibilism. For the compatibilist, however, there is still room for punishment for the sake of deterrence and rehabilitation, concepts that are consistent with the notion of external influences on action; in fact, they depend on it.

Actually, insofar as political liberals tend to favor deterrence and rehabilitation over punishment for its own sake, the arguments for compatibilism at the same time constitute an argument for liberal positions on crime and punishment. The same could be said for liberal attitudes toward equalization of economic and educational opportunity: All these stress the impact of history and circumstance on both choices and outcomes.

The political conservative typically replies, at this point, that society cannot function without people having a strong sense of responsibility over their own lives. You can't just have everyone blaming his or her past for everything he or she does wrong. Society is better off if everyone believes in a stronger sort of freedom, even if that sort of freedom is logically incoherent. This actually seems plausible, which raises a disturbing question: Could it be that it is more important to preserve a misconception about freedom than to know the truth? Are there other areas where this principle applies—where a misconception is more useful than the truth? Perhaps it is better for everyone to believe in objective moral rules, even if a philosophical justification for such rules is not available. Maybe it would be better if everyone believed in a god who will punish misbehavior, even if we have no legitimate evidence for the existence of such a being.

WORKS CITED IN THIS CHAPTER

Aristotle. *On Interpretation.*

Chrysippus. www.theinformationphilosopher.com.

Leibniz, G. W. *Theodicy.*

MacFarlane, John. "Future Contingents and Relative Truth," *Philosophical Quarterly* 53 (2003), 321–336.

Smith, Quentin, and L. Nathan Oaklander. *Time, Change, and Freedom: An Introduction to Metaphysics* (New York: Routledge, 1995).

Other works of relevance to the issues in this chapter

Earman, John. *A Primer on Determinism* (Dordrecht, NL: Reidel, 1986).

Hasker, William. *God, Time, and Knowledge* (Ithaca, NY: Cornell University Press, 1998).

Sobel, Jordan Howard. *Puzzles for the Will* (Toronto: University of Toronto, 1998).

Williams, Clifford. *Free Will and Determinism: A Dialogue* (Indianapolis, IN: Hackett, 1980).

Chapter 8

Could the Universe Have No
Beginning or End in Time?

Did time itself begin, and will it end? Has the universe always been around, and will it one day cease to exist? Two very different research methodologies have been applied to addressing such issues. Relationists (represented by Aristotle and Leibniz) and idealists (represented by Kant) have each tried to use reason alone to work out the possibility of a beginning or end to time. Realists (represented by modern physical cosmologists) rely on empirical methodology, albeit supplemented by a lot of conjecture, in arriving at some ideas on the subject.

RELATIONISM: ARISTOTLE AND
ETERNAL CHANGE

Aristotle thought that his analysis of time put him in a position to draw conclusions about the permanence of the universe. He denied that there can be a beginning or end to time. Recall that, for him, time is just the measure of change. He argued that, although there can be a first moment of a particular process of change, there can be no ultimate beginning or end to change itself. For any change to take place, the conditions of change have to be in place, and that requires a prior process that brought the conditions of change into place. Similarly for an end to change: Any end to a process of change itself represents a process the end of which requires a further change-ending event, and so on. Hence there cannot be a beginning or end to time because there cannot be a first or last change, and time is just the measure of change. Both time and the changing universe that it measures are eternal. Any moment of time, for Aristotle, can only be a midpoint between an earlier and a later time.

It is indeed hard to accept the notion of a beginning to time. Wouldn't a beginning to the universe have to be preceded by something that gives rise to it? But then this so-called beginning would not be the beginning of time itself: It would just be something that happens in time.

It is interesting to note that history's other great temporal relationist, Leibniz, came to the opposite conclusion from Aristotle's. Like Aristotle, Leibniz believed that time is just a relation—an intellectual abstraction we employ in understanding the real. Motivated by a theological perspective that demands a beginning to (i.e., the creation of) the world, he argued (in a letter that would later become part of the *Leibniz-Clarke Correspondence*) that the universe, and thus time itself, began with the first event or state of affairs. Unlike Aristotle, he did not feel that a beginning to time implied a nonsensical time before time. He did not contrast his position explicitly with Aristotle's, but he seemed to feel that Aristotle's reasoning treated time too much like a Newtonian container for events, rather than as an abstraction from causal relations. If the universe begins, then time does; but this does not imply a temporal location for the beginning of time.

In the *Correspondence*, Newton's ally, Samuel Clarke, complains that Leibniz can't have it both ways: You can't say that the universe was created, but not created *at some time*. Clarke also claims that Leibniz limits God's abilities. For the relationist, time is created when the universe is, because time is just a function of events. But then God would not have been able to create the universe at a different time, which one would think would be an option for an omnipotent creator. Leibniz responds that there is a way for the universe to have been around longer: It would have been around longer, in the relationist sense of the word, had God simply put more events into it. By way of explanation, he supplies his own diagram in figure 8.1.

The actual universe begins with events at A-B-C-D; had God wanted more to happen, He could have created some events that led to A-B-C-D. Events at R-S-A-B are causally related to events at A-B-C-D, and temporal succession is just what we call the relation between non-coexisting, causally related events. Creating a universe with more causally related events in it would amount to creating the universe 'earlier.' Thus, Leibniz claims, the relationist can have a creation story without an absolute time at which the creation takes place.

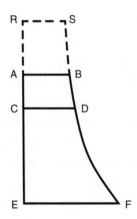

Figure 8.1

IDEALISM: KANT'S DICHOTOMY

The temporal idealist Kant rejected relationism because we unavoidably think of time as a quantity—something there can be more or less of—rather than as a kind of relation. The relationist notion of having more time only in the sense of having more changes or sequentially related events, he felt, just did not capture the role it plays in our sensibility. Although he was not a realist like Newton, Kant did agree that we intuitively *represent* time as a sort of Newtonian container for events, which inevitably gives rise to the question of whether the container is infinite or not.

He addresses the issue of the temporal limits of the universe in his characteristically clever fashion. The section of the *Critique of Pure Reason* in which he tackles the problem is called the first "antinomy of pure reason." For Kant, an **antinomy** is a situation in which you are confronted with two logically valid arguments for contradictory conclusions. Any situation like this would entail that one or both arguments had a false premise. He thought that, given certain common (but false) presuppositions, one could come up with a logically valid argument both for and against the conclusion that the universe is unbounded in time. Exposing the contradiction is intended to point the way to dispensing with those false presuppositions that make the antinomy possible.

He presents each of these contrary arguments before showing how to resolve the apparent contradiction. First, he offers what he considers a valid proof that the world must have had a beginning in time. "For if one assumes," he begins, "that the world has no beginning in time, then up to every given point in time an eternity has elapsed, and hence an infinite series of states of the world, each following another, has passed away." If the world has no beginning in time, then at any point an infinite number of hours must have actually elapsed. But an infinite series of finite time periods cannot actually be completed. If an infinite amount of distance has to be crossed to get someplace, then that place cannot be reached; if an infinite amount of time has to pass for something to happen, then it never will happen. We cannot now be experiencing a time that sits at the end of an eternity, because eternity doesn't end.* Kant claims that this line of reasoning conclusively shows that the world can only have been around for a finite length of time. (Kant was not thinking in terms of the static theory of time. We shall ignore, for the sake of argument, what effect the static theory might be thought to have on this line of reasoning.)

Kant contrasts his argument for the finiteness of the world in time with another allegedly drop-dead argument for the opposite conclusion. This time, he asks us to suppose that the universe *has* a beginning in time. Just as Aristotle had said, there seems to be an intractable conceptual incoherency in the notion of a first time. How can you have a moment without a moment preceding it? This is just as absurd as supposing an edge, or boundary, to space: How can you have a spatial boundary without a space on the other side of it? Bounded time would mean that

> there must be a preceding time in which the world was not, i.e., an empty time. But now no arising of any sort of thing is possible in an empty time, because no part of such a time has, in itself, prior to another part, any distinguishing condition of its existence rather than its non-existence (whether one assumes that it comes to be of itself or through another cause). (Guyer/ Wood, trans.)

---

* This argument echoes eleventh-century Persian/Arabic philosopher al Ghazali's argument against infinite past time.

Kant notes that a time before any universe or events in the universe—an "empty time"—would be a time with no distinguishing features. If literally nothing is going on, then what would explain why the universe would come into existence, or the first event occur, at *that* very moment, rather than at some other empty time?* This line of reasoning, Kant claims, appears to conclusively show that the universe can have had no beginning in time.

This pair of arguments leading to contrary conclusions leaves us with a dilemma that must be resolved. Both arguments (according to Kant, anyway) are logically valid, which means that either's failure must be due to a false premise. In fact, Kant argues, they each share the same false premise: namely, that time is something real in itself. If time were something real in itself, then there would be a fact of the matter as to whether the universe is infinite in time. His solution to the conundrum about the extent of time is just his idealism about time. So the answer, for Kant, is "none of the above": Time is neither an infinite nor a finite quantity because it is not a quantity. The problem with arguing about the extent of time is that time does not have an extent; time is just the form in which we experience events, rather than a real bounded or unbounded container for events.

REALISM: BIG BANG PHYSICS

Of course, relationist and idealist conclusions about the extent of time have only as much appeal as the relationist and idealist analyses

* This question, which would be familiar to both Parmenides and Augustine, is intended to apply to both the creationist and the non-theist. In his parenthetical remark in the above passage, he observes that this would be a problem if the universe "comes to be of itself," because there would be no cause for things to happen at any particular moment of empty time. This is similarly so if the universe comes to be "through another cause": A divine creator would have no reason to act at any particular moment of empty time over any other.

of time do. As we have seen, space-time realism is a presupposition in relativistic physics—admittedly, with a number of substantial caveats having to do with the incomplete state of physics, along with deep issues regarding its goals vis-à-vis describing reality as it is in itself. To the extent that science claims to capture reality, what can contemporary cosmologists say with confidence about the origin and fate of the space-time continuum?

Thanks to the work of astronomer Edwin Hubble and others beginning in the early twentieth century, it has been firmly established that the universe is expanding; further, its expansion can be traced back to a highly compressed state that experienced a rapid expansion (about fourteen billion years ago) called the "Big Bang." Was this a true 'beginning' of the universe and of space-time? If so, can this version of a beginning to time avoid the objections presented by Aristotle and Kant? Finally, what does our best grasp of cosmology have to say about the ultimate fate of the universe?

As Brian Greene recently explained in *The Fabric of the Cosmos*, cosmologists appear to be getting closer to a verifiable theory about what happened at the earliest moments of our universe. Until recently, the Big Bang itself was beyond physical explanation, because relativistic physics does not have the methodological resources to describe matter and energy in such a highly compressed and energetic state: Matter and energy in that state do not exhibit the same forces that relativity is designed to explain. However, over the last few decades, something close to a consensus has developed that the universe's original expansion, along with the observed acceleration in its expansion, could be explained by reference to a kind of field underlying and suffusing all of space-time—the **Higgs field**. The (as of yet) hypothetical Higgs field would have a number of important properties; one of them is its involvement in the existence of a repulsive gravitational force. This repulsive force, according to experts in the field of inflationary cosmology, would explain the Big Bang and expansion of the universe, as well as the distribution of matter and energy in the universe as we know it.

The Higgs field has not yet been observed, but physicists predict that if this field is present, a particle called the Higgs boson will become detectable in very high-energy collisions between other elementary particles; this would then be an indicator that the theory is correct. The experimental detection of the Higgs field is a primary purpose of the new Large Hadron Collider (LHC) in Switzerland. Physicists anxiously await many new discoveries thanks to the collider's operation, but there is a lot of confidence that the Higgs field is real, and that its existence will eventually be confirmed once collisions with sufficient energy have been generated.*

If all this works out, we would have a much better grasp of the initial conditions of the universe, as well as the reasons why space-time expanded and continues to do so. But it is not clear whether this information would do anything to address questions that Aristotle and Kant raised about the conjunction of temporal realism with the thesis that the universe had a beginning. If the Big Bang was the beginning of space-time, then space-time *had* a beginning. To make sense, this needs to be thought of as a beginning *to* time rather than a beginning *in* time: One lesson learned from Aristotle is that a beginning of time in time doesn't make any sense. It remains to be seen, though, whether we can make sense of a beginning *to* time.

One account that would avoid this problem is a version of the **multiverse theory**, where our universe is a development of, or from, another one inaccessible to us. The multiverse model makes our universe a kind of emanation from, or branch of, a much larger system. The notion of other parallel realities has cropped up over

---

* In fact, researchers at the LHC announced a likely Higgs sighting in July 2012!

the last few decades in a number of subfields of theoretical physics. This sort of idea has been proposed to solve a variety of problems. In particular (as physicist Sean Carroll has argued), it might be employed to account for the otherwise wildly unlikely low-entropy state of our early universe.

Multiverse theory is pretty depressing: In addition to reminding us of our own seeming insignificance in a vast universe, this theory makes our entire universe just a kind of multiverse belch. More to the point, the multiverse idea is both implausible and unhelpful when it comes to understanding time. First, the proposed existence of these other universes has no observational consequences, and the principle of parsimony in science (also known as 'Occam's razor') suggests that we should not posit any entities or phenomena not necessary to explain what we observe. Unnecessarily positing an infinity of other universes appears to constitute the ultimate violation of this principle, until and unless the multiverse is shown to be the best explanation for some data before us or shown to be strongly implied by a well-confirmed theory.* Second, it is not clear—at least in the context of this discussion—that the notion of 'another' universe makes any sense: If it exists, is it not part of the universe by definition? There are other, technical contexts in which the notion of multiple universes may be useful, but it remains to be seen whether any such context would be pertinent to the fundamental philosophical question about time. Any philosophical questions about the extent of the universe could just get relocated to the multiverse. We could still ask whether the *multiverse* had a beginning or not, and whether this means time had a beginning.

An increasingly popular theory about the origin of the universe has been proposed by physicists like Paul Davies and Alan Guth:

---

* See Brian Greene's 2011 book *The Hidden Reality* for an extensive discussion and defense of multiverse approaches.

namely, that space-time emerged spontaneously and randomly out of, literally, nothing. (The subsequent expansion of the universe and creation of matter would then proceed according to widely accepted principles.) This answer may seem even less satisfactory than the multiverse theory for several reasons. In particular, science is in the business of talking about the rules upon which one sort of event follows another; yet this theory of the generation of the universe has that momentous event following nothing whatsoever.

There are a few points, however, that make this more than just a totally capricious explanation. Davies points out that more and more space-time is literally being created from nothing all the time as the universe expands. And quantum physics allows for genuinely random phenomena—even the creation of elementary particles out of nothing and for no reason. So the notion of the universe as "the ultimate free lunch" (Guth's phrase) is not as unfounded as it might sound.

Even so, it's not clear that this idea of a universe via spontaneous generation addresses Kant's conceptual problem of a boundary to time. While the multiverse theory leaves us with time stretching back forever, the spontaneous-generation theory leaves us with a first moment to time. Either option raises metaphysical problems; this dilemma was the motivation for Kant's idealist solution.

In *A Brief History of Time*, Hawking proposes a way of thinking about space-time that may partially address philosophical problems about having to choose between placing or not placing temporal boundaries on the universe. Like Kant, Hawking is troubled by both the proposition that the universe has always existed and the proposition that it had a beginning. A standard relativistic model of the universe, as we have noted, traces its history back to a sort of ultimate compression of all matter and energy. Physics as we know it can say little or nothing about such a state of things. This model of the history of the universe thus yields a result that presents a seemingly intractable philosophical and physical problem: The universe has a beginning in time with no preceding time, and physics cannot explain why the universe comes into being. In the absence of a different approach, neither philosophy nor physics seems able to describe, explain, or otherwise make sense of the origin of the universe.

Hawking answers with a model that describes the universe neither as infinite nor as bounded in time, called the **no-boundary proposal**. He claims that one can represent the universe's history, and the history of space-time, in such a way that it has a beginning and an end but no boundaries (figure 8.2). His favorite analogy compares the beginning of the universe to the Earth's North Pole and the end to the South Pole. The circumference of the globe at any given latitude represents the spatial extent of the universe. The universe is bigger in the 'middle' of its history and shrinks to a point at its ends.*

The key achievement of this model is that, in it, space-time has a beginning and end, without itself being bounded by anything. The North and South Poles are the 'beginning' and 'end' of the planet, but they are not edges or boundaries; though Hawking's

---

* From *A Brief History of Time* by Stephen W. Hawking, ©1988, ©1996 by Stephen W. Hawking. (Used by permission of Bantam Books, a division of Random House, Inc. Any third-party use of this material, outside of this publication, is prohibited. Interested parties must apply directly to Random House, Inc., for permission.) This model includes a 'big crunch' where, at some point in time, the universe reverses its expansion and collapses in on itself—thus the need for a 'South Pole' in the analogy. This was less clear at the time he was writing the book, but very recent (starting in 1998) astronomical observations increasingly point to an *acceleration* in the universe's expansion. The most likely prediction, at this point, is of an ever-expanding universe that eventually reaches a state of maximal distribution of matter and dissipation of energy. In other words, the universe will eventually enter into an utterly cold and inert state, forever. If so, the 'back end' of space-time would look to be unbounded in the sense of being infinite.

THE EARTH

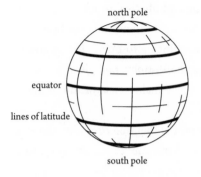

THE UNIVERSE

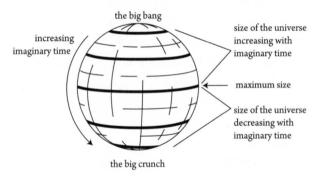

Figure 8.2. Hawking's speculative representation of a temporally self-contained universe.

model represents the universe as finite, it does not require a 'time before time' any more than our globe requires anything on Earth to be located north of the North Pole. Further, the usual laws of physics apply at the North and South Poles, just like anywhere else. According to this model, space-time has, in a sense, a 'first' point, without requiring an edge that implies something lying on the other side.

But the Big Bang, one might reply, is an event. How can any event, even the Big Bang, not happen in time? And if it happens in time, then there must be a time preceding that moment. The answer is that the Big Bang is not an event. From philosopher of religion Bruce Reichenbach:

> [G]iven the Grand Theory of Relativity, the Big Bang is not an event at all. An event takes place within a space-time context. But the Big Bang has no space-time context; there is neither time prior to the Big Bang nor a space in which the Big Bang occurs. Hence, the Big Bang cannot be considered as a physical event occurring at a moment of time. As Hawking notes, the finite universe has no space-time boundaries and hence lacks singularity and a beginning (Hawking 116, 136). Time might be multi-dimensional or imaginary, in which case one asymptotically approaches a beginning singularity but never reaches it. And without a beginning the universe requires no cause. The best one can say is that the universe is finite with respect to the past, not that it was an event with a beginning.

Hawking explains how the static theorist's block model of the universe naturally suggests this approach:

> In general relativity, time is just a coordinate that labels events in the universe. It does not have any meaning outside the space-time manifold. To ask what happened before the universe began is like asking for a point on the Earth at 91 north latitude; it is just not defined. Instead of talking about the universe being created, and maybe coming to an end, one should just say: The universe is.

Thus we would have a universe with an earliest stage *of* real space-time, but no beginning *in* time. This model, then, represents an alternative to relationism or idealism that appears to neatly address our concern about the extent of time.*

CONFRONTING OUR OWN LIMITATIONS

The no-boundary model, though it neatly evades the old philosophical conundrum about the extent of time, seems to suggest something about the universe that many have found deeply unacceptable: *namely, that there is no prior external reason why the universe is the way it is or why it exists at all.* The no-boundary proposal doesn't require a prior explanation of the existence or nature of the universe: The universe just is, and it just is the way it is. Nothing in this proposal brings about the beginning of time; nor is there any particular demand in the model for an explanation as to why the laws of nature are the way they are. The theological significance of this idea is not lost on Hawking:

> If there is no boundary to space-time, there is no need to specify the behavior at the boundary—no need to know the initial state of the universe. There is no edge of space-time at which we would have to appeal to God or some new law to set the boundary conditions for space-time. We could say: "The boundary condition of the universe is that it has no boundary." The universe would be completely self-contained and not affected by anything outside itself. It would neither be created nor destroyed. It would just BE. As long as we believed that [the] universe had a beginning, the role of a creator seemed clear. But if the universe is really self-contained,

---

* The point of all this is not that the no-boundary model is correct, but that it shows one can at least coherently describe a fully realized model of the universe, including a real space-time, that requires no cause of the universe's existence lying either in or outside of time.

having no boundary or edge, having neither beginning nor end,
then the answer is not so obvious: What is the role of a creator?

Needless to say, theologians don't much care for this result. What
argument can they make against it? Hawking's proposal seems to violate a principle that some theologians, and even some philosophers,
have historically claimed as a necessary truth: that the existence of
every thing has a cause of its existence, and that every positive fact
has an explanation or reason for its truth. Belief in this principle (we
will follow Leibniz in calling it the **Principle of Sufficient Reason**,
or PSR) is usually tied to belief in a divine creator: PSR is the basis
for a well-known line of reasoning purporting to prove the existence of a conscious creator of the universe. Arguments of this sort
are known as **cosmological arguments** for the existence of God.
Notable proponents of some version of a cosmological argument
include Aristotle, eleventh-century Persian/Arabic philosophers Ibn
Sīnā and al Ghazali, thirteenth-century Catholic theologian Thomas
Aquinas, and Leibniz. It remains popular today among many theists,
and it might even be described as fundamental to the Abrahamic
religious traditions. The argument comes in two main versions,
each of which depends on PSR: the First Cause Argument, and the
Argument from Contingency. These two versions of the argument
correspond, respectively, to the two options considered by Kant:
that the universe either has a beginning in time, or that it has always
existed. The First Cause Argument presumes that the universe had
a beginning in time, and claims that there must therefore be a first
cause to commence the chain of events constituting the universe:

The First Cause Argument:
- The universe had a beginning.
- Anything with a beginning has a cause.

- Hence, the universe had a cause.
- The only possible cause for the universe is a divine creator.
- Hence, a divine creator exists.

The Argument from Contingency, in contrast, is designed to avoid having to assume the universe had a beginning in time:

The Argument from Contingency:
- Anything that exists contingently (i.e., does not exist by logical necessity) must have a reason or explanation for its existence.
- The universe exists contingently.
- Hence, the universe has a reason or explanation for its existence.
- The only possible reason or explanation for the existence of the universe involves a divine creator.
- Hence, a divine creator exists.

The first argument nicely sums up a theistic version of the rejection of the idea of bounded time without prior cause; the second argument represents a theistic rejection of a universe that has simply been around forever without explanation.

The First Cause Argument suffers from several deficiencies: First, it presumes that the universe has a beginning in time; as we have seen, Aristotle and Kant each raise serious doubts that this is even a coherent option. Second, the multiverse theory would answer it with a greater, infinite multiverse with no beginning in time. Third, Hawking's model coherently describes a finite universe with no beginning in time. So it doesn't just go without saying that the universe has a beginning in time. Fourth, the First Cause Argument presumes that everything has a cause; but quantum physics—elements of which, at least, have been decisively confirmed—may allow uncaused events, at least at the level of the properties and behavior of elementary particles.

The Argument from Contingency avoids some of these objections: It does not presume that the universe has a beginning. It only says that, logically, the universe may or may not have existed, and there must be an explanation as to why it does. But why must everything have a reason or explanation? (In other words, why should we think PSR is *true*?) We do tend to find in our everyday experience that things that happen are caused to happen and are therefore explicable. From this we naturally extrapolate to all the cases where events happen and we either don't experience the cause or don't have the information needed to infer back to it; because the facts we do investigate tend to have explanations, we suppose that those we don't or can't investigate have explanations as well.

The presumption that all events are potentially explicable encourages us to form theories about the way the world works (lightning makes fire, buffalo meat is good to eat)—theories with predictive value. This tendency would have enormous adaptive advantages; it is likely that natural selection has favored the instinctive presumption of explicability. But theistic cosmological arguments take a thesis about events or facts *in* the world (namely, that they each have a cause or explanation) and apply it to the very existence of the whole universe. The universe may be thought of as constituted by events or facts; the universe is, one might say, the sum of all events and/or facts. Does it just go without saying that the empirically grounded assumption that everyday events and facts have a cause or explanation applies to the meta-fact that there *are* events and facts? (Kant questioned this very extension of the demand for causal explanation as a misapplication of the concept of causation.) Given our experience, it is understandable that we should find the question "Why is there something rather than nothing?" meaningful. But it is not clear that our experience actually gives us any evidence that this is a legitimate question.

As is often noted by philosophers of science, explanation has to end somewhere. A fact about the world may be explicable by its subsumption under some natural regularity, which is explicable by reference to some more fundamental fact, which is in turn explained by some more fundamental natural law, and so on. But at some point, things just are the way they are. Bertrand Russell called these "brute facts." The only alternative is a chain of explanations without end, which is hardly more satisfying.

The theist doesn't accept brute facts about the natural world; she insists on the pertinence of the question of why things are the way they are—the answer to which, furthermore, requires a divine creator. But this question is motivated by the alleged rule that all facts need an explanation: If that is true, then the fact of a divine creator would seem to require one too. So, on the theist's own premises, the proposed creator cannot function as an ultimate explanation.

The theist's only option, at this point, is to reply that the existence of a divine creator is just a brute fact about the universe with no further explanation. But if that sort of answer is acceptable, then (as mathematician John Allen Paulos has recently proposed) why not stop with the fundamental laws of nature? Why can't they, and the existence of the universe, be the ultimate brute fact? That would be simpler than adding to the story a whole supernatural layer, a massive complication for which we have no good evidence anyway.

Furthermore, as philosopher Robert Nozick once asked, if there were nothing (i.e., if the universe didn't exist), then wouldn't it be just as good a question to ask why there is nothing rather than something? There are many ways for there to be something, but only one way for there to be nothing. That makes nothing a much more special state than something; if there were nothing, then that would seem to call for an explanation all the more.

If there were a God, we could ask why there is one; if there weren't, we could ask why not. It starts to become clear, at this point, that we are capable of articulating questions that are grammatically well formed but literally have no answer. It may be that "Why does the universe exist?" is one of those questions. This is frustrating, but no one promised us that there is an answer to every question we are capable of asking. Our dissatisfaction with unanswerable questions, or with Hawking's finite yet unbounded and unexplained space-time, likely stems from an instinctive expectation based on a natural and adaptive tendency to look for theories about how things around us work. If so, such dissatisfaction has to do more with a psychological resistance to accepting things the way they are than with a lack of information (a common problem; accepting our own mortality is another example). Taking Shakespeare a bit out of context, the fault, perhaps, is not in our stars but in ourselves.

WORKS CITED IN THIS CHAPTER

Alexander, H. G., ed. *The Leibniz-Clarke Correspondence* (Manchester, UK: Manchester University Press, 1956).

Aquinas, Thomas. *Summa Theologica*.

Aristotle, *Physics*.

Davies, Paul. *Superforce* (New York: Touchstone, 1985).

Greene, Brian. *The Fabric of the Cosmos* (New York: Knopf, 2004).

------. *The Hidden Reality* (New York: Knopf, 2011).

Guth, Alan. *The Inflationary Universe* (New York: Basic Books, 1998).

Hawking, Stephen. *A Brief History of Time* (New York: Bantam, 1988).

------. "Quantum Cosmology," in *300 Years of Gravitation*, ed. by Stephen Hawking and Werner Israel (Cambridge, UK: Cambridge University Press, 1987).

Kant, Immanuel. *Critique of Pure Reason*, trans. by Paul Guyer and Allen Wood (Cambridge, UK: Cambridge University Press, 1998).

Nozick, Robert. *Philosophical Explanations* (Cambridge, MA: Harvard University Press, 1983).

Paulos, John Allen. *Irreligion* (New York: Hill and Wang, 2007).

Reichenbach, Bruce. "Cosmological Argument," in *The Stanford Encyclopedia of Philosophy*, ed. by Edward N. Zalta, http://plato.stanford.edu/archives/win2010/entries/cosmological-argument/.

Other works of relevance to the issues in this chapter

Falk, Dan. *In Search of Time: Journeys along a Curious Dimension* (New York: Thomas Dunne Press, 2008).

Mackie, J. L. *The Miracle of Theism: Arguments for and against the Existence of God* (Oxford, UK: Oxford University Press, 1983).

Rowe, William. *Philosophy of Religion: An Introduction* (Belmont, CA: Wadsworth, 2000).

Stenger, Victor. *The Comprehensible Cosmos* (Amherst, MA: Prometheus, 2006).

Is "What Is Time?" the Wrong Question?

We have seen that the discussion concerning the reality of time keeps coming back to Parmenides' question as to whether there is a real distinction between past, present, and future. Israeli philosopher Yuval Dolev thinks that asking this ontological question, to borrow Wittgenstein's phrase, just "replaces an unspecified confusion with a systematic one." Dolev argues that the term 'real' is misapplied in this context; 'real' is usually contrasted with fake or illusory, but that is not the issue here. What we mean by past and future is inextricable from our way of experiencing events. What would be the consequences, after all, of concluding that all times are equally real, and there is nothing objectively special about the present? We still must act as though the present is special, because agency and, indeed, survival are impossible otherwise. He suggests that our goal should be to leave behind the question of the reality of time altogether, in favor of a closer examination of temporal experience.

Recall the coffee drinker thinking of her cup as *here*. Clearly, "here" is not an objective designation for anything. The universe would not recognize anything as either here or there; as far as the universe is concerned, things just are where they are. But for the individual coffee drinker at that particular place and moment, such a thought is a meaningful guide to action. Say that we accept that McTaggart and Einstein are right, and there is no privileged, moving *now* that we occupy and that has special significance. Yet what time it is now matters to me now, even if from an objective perspective it means nothing at all. My beliefs about the here and now, as well as about the (subjective) past and future, explain not only my actions but also my emotions; such beliefs constitute the essence of my basic experience of the world. Clearly, we would be missing something important if we totally ignored these perspectival assessments in favor of what things look like from the impersonal universe's standpoint. For one thing, we would be unable to live.

This tension between what is objectively the case and what is subjectively significant is reminiscent of the existentialists' conclusions in light of their belief that the universe is godless and inherently valueless. Existentialists like Nietzsche, Camus, and Sartre characteristically hold this belief, yet insist that our choices, moral and otherwise, with regard to what we find important do matter. Even though they also characteristically hold that freedom is itself an illusion, they maintain that our choices imbue the universe with value. True, this means that what one values only has a conditional, subjective worth, but is that less meaningful to the valuer than any hypothetical objective values that he or she might not have shared anyway? Our lives are of no concern to the universe, but they are of immense concern to ourselves … and isn't that enough to go on?

There is, then, an argument that we need not concern ourselves with the conclusions about time coming out of philosophy and physics: Our experience of the world is real and fundamentally involves dynamic temporality and temporal directionality. Whether or not a theoretical objective perspective would include change and direction is irrelevant to our human concerns.

It is indisputable that the conclusions we have already reached about time do not mean that we should (or even can) abandon our ordinary ways of thinking about time, given that our experience of it is inherent to perceptual and emotional awareness. Still, I wouldn't go quite so far as to conclude that our ongoing inquiry into the nature of time is totally off-track. The scientific study of nature truly improves our grasp of time, even if it doesn't decide the issue of whether we are supposed to take the present seriously. Further, when you have a scientific research program, as we do in the case of the scientific investigation of space-time, perception, and human psychology, there is need for philosophical reflection on its motivation, meaning, and results.

Building on what we have learned from the history of the philosophy of time, I would propose the following: If an answer to the question "What is time?" still seems to elude us, perhaps it is because we have been asking the wrong question. *Time is not so much a 'what' as a 'how,' and not so much a question as an answer.*

Time as we know it in experience is a matter of how we adaptively organize our own experiences; in a physical and cosmological context, it is a matter of how we can most successfully model the universe of occurrences. As such, time is an answer: a solution to the problem of organizing experience and modeling events.

So who is right, the relationist, the idealist, or the realist? The answer lies partly in seeing that each position has something to be said for it.

Relationists have a point in that much of what we have to say about time has to do with our mode of organizing and relating events. In that sense, you could call time a kind of relation. The measurement of time is possible only in terms of observed motions or changes, such as the orbit of the Earth. It is for this reason, as relationist P. J. Zwart points out, that we say (albeit only metaphorically) that "time stands still" in a place where nothing changes.

Idealists are right in that our grasp of time will always be mediated by our way of understanding things. Temporal experience is a kind of construction, rather than a mere reflection of nature. We can never penetrate to the sheer, naked reality of things as they are in themselves, unmediated by the conditions under which we experience things (see figure E.1).

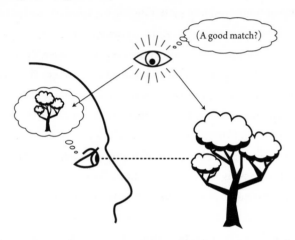

Figure E.1. A good match between representation and reality? This is a perspective we can never achieve, and thus a comparison we can never make.

Whatever we come up with as a description of nature will always represent a particular way of understanding nature and never a final, unique, fully independent description. There is no way for us to step outside ourselves as a species and directly compare our representation of nature with nature in itself, in order to see if the former is an accurate reflection of the latter.

Realists, however, get support from the fact that there are, objectively, more and less successful models of reality. One of the best illustrations of this principle is the Michelson-Morley experiment, described in chapter 3, and the way in which it pointed to the superiority of the theory of relativity over Newtonian absolute space and time. Einstein-Minkowski space-time is part of a demonstrably superior representation of reality: It explains more and allows better predictions (even as physicists work on an even better theory). The goal of the physical scientist is to find the most comprehensive and effective theory of nature. A theory that treats space-time as real exhibits a good 'fit' with observation in the limited sense that it is simpler while yielding wide-ranging explanatory and predictive power. Given that the supposition of a real space-time gives us just such an elegant and useful theory, is it inappropriate to declare that space-time is, indeed, real?

It is true that we cannot independently verify the fit of a theory with reality itself. We are constituted to experience the world in a certain way. Any means we could come up with to verify the accuracy of the way we experience the world would depend on the very same innate experiential and conceptual scheme we are trying to validate. Yet theories that have done well in terms of explanatory power while maintaining simplicity have the best track record with regard to subsequent discoveries. It is not a guarantee, but the scientific method has served us well over the last few centuries. It is true that we can never know the universe or its laws independently of the conditions under which we can experience them. Yet the guidance we are able to glean from reality—via the experimentally confirmed replacement of inferior models of nature with superior ones—means that, despite inescapable limitations, we can hope to move closer to the truth. This fact makes some kind of realism about time, and a continued commitment to related ontological questions, defensible and even fruitful.

Regarding the nature of time itself, results in logic and physics entitle us to speak authoritatively in terms of a static space-time continuum. At the same time, philosophy (by examining the necessary conceptual presuppositions of experience), together with various empirical studies (such as neuroscience, psychology, evolutionary studies, and even social sciences like anthropology), can clarify the nature of, and limitations on, our experience of time. These studies can validate our projective and irreducible *experiencing* of time as dynamic, as long as such validation is understood in the proper context.

The philosophical study of time and time awareness has yielded other substantive achievements. Zeno's challenges, for example, help us clarify what we mean by change and motion. Philosophical questions about temporal experience show us what to look for as we study time perception in the brain. The logical analysis of the dynamic theory of time bolsters the physical case for the static theory and helps us distinguish between real phenomena with respect to time awareness versus mere conceptual presuppositions; it also leads to important hypotheses about time awareness for evolutionary theory to pursue. The philosophical analysis of proposed explanations of time's directionality tells us what would or would not count as a scientific solution. A philosophical examination of the logical possibility of altering the past, in combination with a physical understanding of space-time, tells us what sort of time travel is possible. Chrysippus' conceptual analysis of freedom, together with an appreciation of the static theory of time, shows us what free will can (and cannot) amount to; this discussion, furthermore, needs to be taken into account in the application of the concepts of responsibility and justice in social contexts.

Philosophy accomplishes the most when it works hand-in-hand with the empirical sciences. The field of time studies is a terrific example of this partnership. Philosophical analysis clarifies the questions that need to be addressed in this area, and it is essential to understanding the significance of the results. The study of time is difficult, but we are not just spinning our wheels. The history of the philosophy of time is a history of progress—progress on a subject that could hardly be more elemental and momentous.

INDEX